MW01286940

Lumberjack

INSIDE AN ERA IN THE UPPER PENINSULA OF MICHIGAN

By William S. Crowe

EDITORS:
Lynn McGlothlin Emerick
Ann McGlothlin Weller

FIFTIETH ANNIVERSARY EDITION

ISBN 096-50577-3-9
Library of Congress Control Number 2002104698

Published by North Country Publishing
355 Heidtman Road • Skandia, MI 49885
(906) 942-7898
Toll free: 1-866-942-7898
northco@up.net

Printed in the United States of America
Photographic Reproduction—Photomaster, Marquette, MI
and the State Archives of Michigan, Lansing, MI
Book and Cover Design—Holly Miller, Salt River Graphics, Shepherd, MI
Illustrations—Carolyn Damstra, East Lansing, MI
10 9 8 7 6 5 4 3 2 1

In memory of

William Scott Crowe, our grandfather, who lived it,

Helen Crowe McGlothlin, our mother, who kept her promise,

and for Stephen, Lynn, Mary and Susan,

the next generation

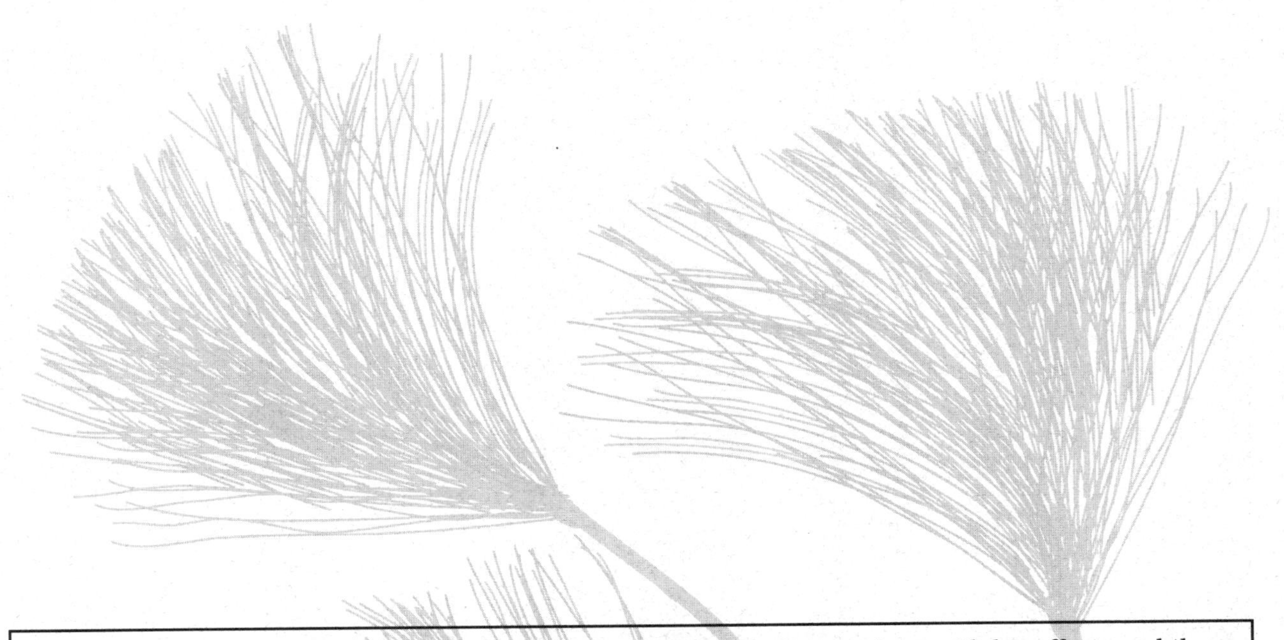

The stories and experiences of our grandfather, William S. Crowe, along with his efforts and those of our mother to create the First and Second Editions of *Lumberjack*, have long been a part of our remembrances.

The Golden Anniversary year of the original publication of these stories seemed an appropriate time to bring *Lumberjack* back into circulation. Crowe's experiences over a century ago in the north woods of the Upper Peninsula of Michigan provide a vivid link to those times for a new generation of readers.

Contents

PREFACE

The History of Lumberjack—Inside an Era

In his Introduction to the First Edition, published in 1952, the author, William S. Crowe, explained how *Lumberjack* came to be:

In the fall of 1947 several letters appeared in the Escanaba Daily Press *speculating on the identity and meaning of certain "mysterious" marks found on lumber used in construction of the Schoolcraft County Memorial Hospital at Manistique, various theories being advanced ranging from Egyptian hieroglyphics to the American Indians. The lumber had been sawed from "deadhead" logs that had been submerged in the Manistique River for fifty years or more, and then salvaged by a local sawmill.*

It seemed incredible to me that there should be any mystery about the origin and purpose of these marks, as I supposed everyone was familiar with the system of log marking used in the old pine lumber days. I suddenly realized that a colorful era which seems only yesterday to me ended half a century ago and that few of the present generation have ever seen a big sawmill, or even a really big pine tree.

Furthermore, this half century of American history witnessed changes of kaleidoscopic rapidity far beyond any similar period prior to the American Revolution.

The marks in question were very familiar to me because part of my job with the Chicago Lumbering Company was to keep a tally of them, so I consented to write a few articles descriptive of those times.

Here and there some digressions into the general field of economics, which some may think foreign to the title, are only intended to give a clearer overall perspective of life in those days. It would be hard to understand the lumberjacks and their actions and ways of life without some knowledge of the background and a general picture of actual life at that time.

However, I have tried to confine my digressions to only those phases that have a direct bearing on the life of the lumberjack and the White Pine era.

I am not a professional writer, and if the personal pronoun crops up too often, my apology is that this is simply a story told in as simple language as possible about matters and events in which I was either an active participant or a firsthand observer. This gave me one advantage over certain professional writers who depended largely on hearsay. I was able to screen out the "chaff" from the Bunyanesque tales with which the lumberjacks delighted to regale their listeners, and I can therefore guarantee the reliability of what little hearsay does appear in these articles.

Wm S Crowe

Wm. S. Crowe

FIRST EDITION—1952

Crowe's articles, published in the *Escanaba Daily Press* and the *Manistique Pioneer-Tribune* at various times during the 1930s and 1940s, aroused much interest in the history of the area and region and he was urged to publish a book of his experiences. *One day, Mr. Harold "Rabb" Klagstad, a former Manistique resident and partner in and head of the Chicago branch of Ernst and Ernst (a nationally known accounting firm), asked me to put these articles together in a book. I told him that writing a book and publishing a few letters of local interest were vastly different propositions but he said if I would give him copies of the letters, his firm would print the book, provided I would give him copies to distribute to colleagues and friends. That is how the book came to be written.*[1] The very small print run of Lumberjack soon sold out.

SECOND EDITION—1977

The response to the First Edition of *Lumberjack*, and the many requests for copies which could not be filled, encouraged Crowe to begin preparations for an expanded Second Edition. He collected photographs, conducted research, planned trips to the West Coast and contacted key informants throughout the next decade.

I can now give attention to completing in real earnest the second edition of the booklet, Lumberjack. . . . *it will be a book of reference, and historical fact, but interspersed with anecdotes and personal stories so as to make it interesting as general reading for the public.*

The purpose will be to make it a real picture of actual life in those days with mention of many things considered too unimportant by professional writers but which together give a mental picture of the way people actually lived in those days. The new edition will contain everything in the first edition and a lot more in the way of pictures, maps, and detailed information about 100 million feet of "big pine" every year from the "stump" to the decks of the lumber schooners and lake barges for shipment to Tonawanda, Chicago and other ports. The object is not only to give a true picture of the lumberjack and his habits, but also to recreate the atmosphere of those old days and the background of the times in which he lived.[2]

In September 1965, William Crowe fell ill and died six weeks later. In an article upon publication of the Second Edition, his daughter, Helen Crowe McGlothlin, said, "From his hospital bed, he looked at me and said, 'You'll get my book published, won't you?' And I said, 'Well, yes.'"

It took her twelve years to keep her promise, but in 1977, twenty-five years after the original book, an expanded version of *Lumberjack* was released. Ted Bays of Bay de Noc Community College edited Crowe's thirty-two articles into eight chapters; Helen McGlothlin added many new photographs from Crowe's collection. The Second Edition was printed and published by Frank Senger of Senger Publishing Company, Manistique; the Second Edition has also long been out of print.

---

[1] WSC letter to Mrs. Clifford Crick, July 7, 1961.

[2] WSC preliminary memo for letter to Timothy Pfeiffer, November 22, 1957; general memo, October 1, 1962; letter to Miss Dorothy V. Weston, May 20, 1963; and letter to L. G. Sorden, July 13, 1963.

THIRD EDITION: 2002

As co-editors of this new Fiftieth Anniversary edition, we have honored William Crowe's story as he lived it and wrote it.

Where new information has been included in the text, it has come directly from his own letters, notes and published writings. Where clarifications or expansion of information have been made, they are provided in an Editors' Notes section and referenced by number within Crowe's text. Terms of lumbering and social custom used in *Lumberjack* are defined in an expanded, illustrated Glossary. The sources from which relevant information was obtained are referenced in these two sections.

More than forty historic photographs of life and logging in the white pine days have been added to the Third Edition. As in the Second Edition, some chapter titles and subtitles have been changed or added to reflect more accurately the content within.

Friends, relatives, new acquaintances and staff members of various organizations have been generous in their encouragement, searching for information, dates, historic photographs and little-known or forgotten facts in answer to questions and missing pieces of the puzzle. An Acknowledgments section recognizes those who have assisted with this and previous editions of *Lumberjack*.

William S. Crowe "lived in interesting times" and his life story highlights the building of a country as well as the remarkable experiences of many young men and women of his era. From his papers and those of his extended family, we have developed a biography of William S. Crowe, with selected photographs, and included it in this edition.

Our grandfather's stories, and the life he lived so long ago, have come alive again for us as we reread his correspondence, followed his hopes for his writings, researched obscure references and old publications, delved into log marks and river driving, viewed photographs and maps of that era and shepherded this new edition of *Lumberjack* to life.

Lynn (Crowe) McGlothlin Emerick
Emerick Consulting
Marquette, Michigan

Ann McGlothlin Weller
Services for Publishers
Holland, Michigan

May 29, 2002

Wisconsin Historical Society, Image #WHi-3219

Mature eastern white pines on the Menominee Indian reservation. In much of the Northern Great Lakes region—Michigan, Wisconsin and Minnesota—the trees grew close together and often over 100 feet tall. In the Upper Peninsula of Michigan, there are white pine groves remaining in the Estivant Pines Nature Sanctuary, Keweenaw County; Laughing Whitefish State Scenic Site, Alger County; and Sylvania Wilderness Area, Gogebic County.

CHAPTER ONE

Lumbering Towns and Company Men

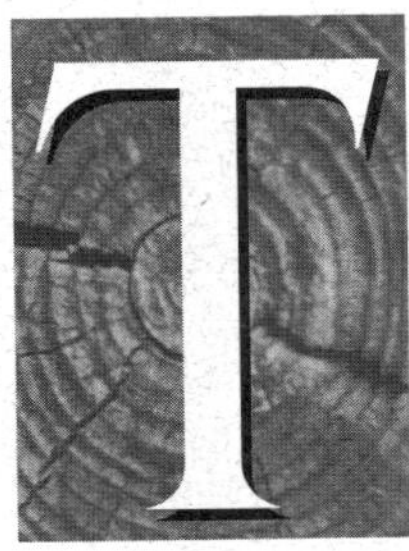

The identity of the mysterious marks found on lumber sawed from deadhead logs was quite clear to me because I was once head bookkeeper for the Chicago Lumbering Company of Michigan (C. L. Co.) and the Weston Lumber Company (W. L. Co.), and have lived in Manistique for over seventy years.

Since Manistique, like many other towns in Michigan during the colorful white pine days, was an almost one hundred percent pure lumber town, the story is mostly about lumberjacks, woods, big sawmills and lake shipping. As bookkeeper for one of the largest lumber companies during the seven years when the industry was at its peak, I had splendid opportunity for firsthand observation.

When it comes to a matter of detail and recreating the atmosphere of those days, the stories of the other lumber towns of Michigan are exactly the same as Manistique; residents lived their daily lives as we did here. The faces changed and the proprietors, the "Lumber Barons," were all different, but lumbering was the same.

"There is no other tree in all the world which has so much of romance . . . as the white pine."

There is no other tree in all the world which has so much of romance, and was so closely associated with people's daily lives and manner of living, as the white pine, which is now almost extinct except for some tracts of closely related sugar pine on the west coast.

And I believe there were no more colorful or interesting or exciting segments of American life than the lumber industries, nor a more picturesque individual than the old-time lumberjack and river driver. They were in direct contact and conflict with Mother Nature in the raw—a comparable industry in this respect being the cattle ranches of the great plains and the western cowboy. I lived for two years on a cattle ranch in Colorado before coming to Michigan so I had good opportunities to get firsthand impressions of both the lumberjack and the western cowpuncher.

A more completely different stage, background and environment could hardly be imagined than the immense pine forests and the treeless, seemingly endless plain, nor two more different individuals than the lumberjack

Log marks from Upper Peninsula lumber companies: (1) Goodenough & Hinds, Delta County, 1889; (2) Garth Lumber Co., Delta County, 1894 (also used by Weston Lumber Co., Schoolcraft County); (3) Charles Mann, Delta County, 1902; (4) J. A. Jamieson & Co., Mackinac County, 1908; (5) Marsh, Koehn & Co., Menominee County, 1904; (6) Oliver Iron Mining Co., Menominee County, 1938; (7) McMillan Bros., Ontonagon County, 1902.

and the cowboy. A cowboy would have been lost almost at once in the great pine forests, and a lumberjack would have been lost even more quickly on rolling treeless plains.

But although environments and ways of life differ, people are fundamentally about the same wherever you find them, and the cowboys and the lumberjacks had the same spirit of daring, resourcefulness, initiative, independence and romance bred by close contact with nature at her best and worst in a vast, free country.

One thing common to cowboy and lumberjack was the similarity of the cattle brands and log marks. The W. L. Co.'s "Barred O" and "Circle O" were duplicated on the cattle ranges, and a big western ranch's "Cobhouse" brand was exactly the same as the C. L. Co's "Cobhouse" log mark.

The quadruple cross # was the familiar C. L. Co. "Cobhouse," and the cross in a circle was the W. L. Co's "Barred O" ⊕. These marks and the Weston Lumber Company's "Circle O" ◎ accounted for about ninety percent of the roughly four billion feet of white pine and red, or Norway, pine these companies cut in their forty-one years of operations, 1872-1912.[1, 2]

Other marks which frequently came up C. L. and W. L. jack ladders were Hall & Buell's HB, the Delta Lumber Company's triangle △, JDW logs cut from J. D. Weston land, Edward Hines & Company SO and the marks of Alger, Smith and Co. Ⓢ, which operated a big mill at Grand Marais supplied by their own railroad to Seney, Germfask and Curtis. The C. L. Co., with a branch office at Seney, operated in that territory at the same time and sometimes cut isolated forties for Alger, Smith

320

LOG MARK
ADOPTED BY

Received for Record, the .... day of .... A. D. 18.... at .... o'clock .... M.
County Clerk.

TO ALL WHOM IT MAY CONCERN:

This is to Certify, That .... of .... do .... in accordance with Act No. 202 of the Session Laws of the State of Michigan for the year 1867, select and adopt the following marks, to be cut or stamped in conspicuous places, upon the ends or sides (or both) of logs to be put by or for .... into .... or .... tributaries for the purpose of floating or rafting to the place of manufacture or shipment.

[L. S.]

Dated at .... this .... day of .... A. D. 18....

MARKS:

For a valuable consideration .... hereby assign and transfer all .... interest in and to the within Log Mark .... to .... hand .... and seal .... this .... day of .... A. D. 18....

Witness, ... [L. S.]

Delta County Clerk; Rod Smith, Marquette

A single company might have more than one hundred log marks. Delta County in the Upper Peninsula received so many registrations that the Log Mark sheet was developed for recording purposes.

and Co. and ran the logs to Manistique in their main river drive.

General Russell A. Alger, President William McKinley's Secretary of War, was the principal stockholder in Alger, Smith and he and Abijah Weston were great friends.

W. S. Crowe collection

Manistique Harbor in the early days.

In the 1880s and 1890s, Hall & Buell had a large sawmill at South Manistique, a community of about 1,200 people on Lake Michigan one mile southwest of Manistique. Not a vestige of South Manistique remains except part of Hall & Buell's old log pond. Their mill was supplied by a railroad from their Indian Lake pull-up and from another pull-up on the Manistique River just above the C. L. Co.'s dump. When the C. L. Co. built the Manistique and Northwestern Railway (now part of the Ann Arbor system) to Steuben and Shingleton in 1896, they bought Hall & Buell's railroad to have access across the Soo Line.[3]

The Delta Lumber Company had a sawmill at Thompson (population about 500 in 1893) supplied by a railroad from a pull-up on the Indian River north of the Big Spring, with branches here and there on the plains between Indian Lake and Cooks west of Delta Junction. A railroad along Lake Michigan connected Thompson with South Manistique and Manistique. The combined population of the three communities in the 1890s was about 5,500.

All told there were over twenty log marks, and a scaler at the head of the jack ladder in each of the five mills entered the scale of every log under its particular mark on a scale sheet. The C. L. and W. L. companies used the combination Doyle-Scribner rule: Doyle for small logs up to twenty-eight inches in diameter, and Scribner for large logs over twenty-eight inches. The mill lumber scale usually was higher than the log scale from ten to thirty percent. In 1913 the lumber scale overran the log scale almost exactly thirty percent.

A NEW DAWN

When I landed at Manistique from the Goodrich Line's *City of Ludington* at exactly midnight, May 29, 1893, I stepped into a strange new world such as I had never seen or even dreamed of. I was only seventeen and had lived for two years on a cattle ranch on the treeless plains of eastern Colorado northeast of Fort Lupton. I had never seen a ship, a large body of water, a sawmill or even a big tree. The screaming saws in five big mills, running twenty-four hours a day; the scent of new lumber and the pine woods; the hoarse whistles of lake steamers; the tall masts of lumber schooners in the harbor; and the flickering flames and red glow from the open burners reflected across the water

Henson Printing Company, Grand Rapids; *Manistique Pioneer-Tribune*

Chicago Lumbering Company saw mills and lumber yards at Manistique.

and in the sky against the dark and somber background of the immense forest—all gave me a feeling early pioneers must have experienced when they discovered a new and unexplored area. I could hardly wait for morning to dawn.

The Chicago Lumbering Company of Michigan was organized in the summer of 1863 by certain Chicago interests who operated a small mill with two circular saws and one gang saw until 1871, when Abijah Weston and Alanson J. Fox came up from Painted Post, New York, and bought the company. From that time until December 1912, when Mr. Lou Yalomstein (then manager of the C. L. Store) and I organized the Consolidated Lumber Company and purchased all their properties, the Chicago Lumbering Company of Michigan and the Weston Lumber Company (they had the same stockholders) were the whole thing in Manistique and Schoolcraft County. The C. L. and W. L. companies owned at this time all of Schoolcraft County, parts of Delta and Mackinac counties and all of Manistique except the wooden saloons in the Flatiron block on Pearl Street, schools, churches and six or seven stores on Cedar and Oak Streets.[4]

Abijah Weston, Fox and their associates were very wealthy men, and when they took over in 1871 things commenced to hum in Manistique.[5] In 1893 when I started to work as time boy, they were

W. S. Crowe collection

The board of directors of the Chicago Lumbering Company of Michigan and Weston Lumber Company. Standing, left to right: William H. Hill, William E. Wheeler, John D. Mesereau, N. P. Wheeler. Seated: George H. Orr, Alanson J. Fox, Abijah Weston, M. H. Quick.

operating three large mills, a planing mill, a charcoal iron furnace and about twenty-six other departments that covered almost every activity found in a community of four thousand people. They employed about twelve hundred men then and I became well acquainted with their various managers and foremen, including: John Quick, C. L. Mill; W. C. Bronson, W. L. Mill No. 1; Sam Mix, W. L. Mill No. 2; John Woodruff, C. L. Lath Mill; E. A. Rose, W. L. No. 2 Lath Mill; R. B. Waddell, Weston Mfg. Co. Planing Mill; H. Duval, Weston Furnace Co.; Arthur DuBois, Manistique Telephone Co.; C. P. Hill, C. L. Store; I. S. Phippeny, W. L. Store; E. W. Miller, Warehouses and Docks; Capt. Lossing, C. L. Dredge; Capt. John McWilliams, Tug *Elmer*; R. P. Foley, Ossawinamakee Hotel; George Wickwire, Retail Lumber; Frank Havilchek, C. L. Machine Shop; A. D. McNair and J. A. Hamill, C. L. Blacksmith; Fred Denzeng, C. L. Harness; Abner Orr, C. L. Barn; M. W. Cutler, W. L. Barn; E. C. Brown, Shipping & Yards; C. J. Thoenen, C. L. Hardware and W. F. Kefauver, C. L. Furniture and Undertaking (on a leaf from the Ossawinamakee Hotel[6] Register dated June 19, 1894, appears the following advertisement: "Chicago Lumbering Co. of Michigan, Undertaking.") The company's furniture and undertaking parlors were located in a building opposite the Elks' Temple, where the Tribune Publishing Co. now stands. H. W. Clarke was cashier of the company's Manistique Bank. Even the cemetery and the post office were company departments for many years, and the village president, clerk and treasurer were usually one or other of the company's officers.

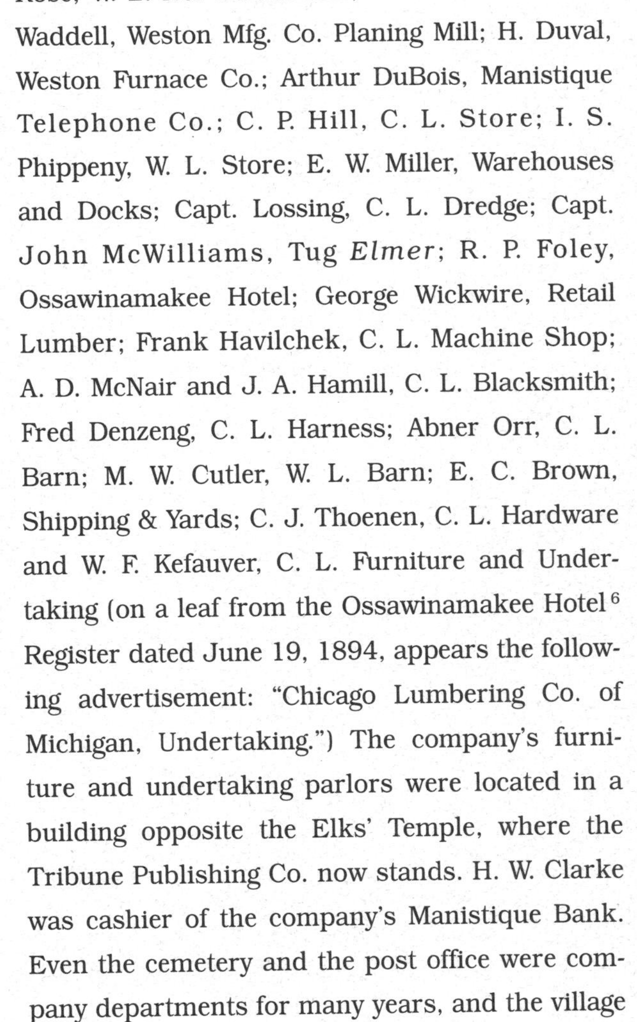

W. S. Crowe collection

The Weston Lumber Company's "new mill," under construction on the west side of the Manistique River in 1884.

J. D. Mersereau, Secretary and Treasurer in charge of the General Office; M. H. Quick, General Superintendent of the Mills and Yards; and George H. Orr, General Woods Superintendent, were all stockholders in both companies. Ed Cookson was walking boss in the woods under Mr. Orr, and prominent camp foremen and jobbers were Frank Cookson, George L. Hovey, "Red Jack" Smith, C. O. Bridges, Paddy Miles, George Scott, Murdoch McNeil, George Roberts, Dugald McGregor, Peter McGregor, William Clemons, the Ferguson brothers of Munising, Bill Lockwood and a number of others.

Abijah Weston, who owned fifty-one percent of the stock in both companies, also had interests in Buffalo, Niagara Falls and elsewhere, including banks, lake shipping and a sizable distributing yard at North Tonawanda, then the largest lumber distributing point in the United States. I estimated his wealth to be somewhere between $30 and $40 million. He died in March 1898.

The combined capital of the two companies in 1894 was $2.1 million (Old Deal dollars, which in terms of today's [late 1940s - early 1950s] depreciated gold dollars and inflated credit money and prices would be about $25 million) and their operations were on a larger scale than most of us today realize. The flagship of the TBL (Tonawanda Barge Line) fleet of lake steamers and barges owned by A. Weston & Son carried about one million feet of lumber, and I used to make out bills for cargo after cargo, not one of them over $15,000; on today's market they would all be worth at least $350,000. In the railroad panic of 1893 the two companies lost approximately $1.5 million, but in one seven-year period with which I am familiar (1893-1900) they paid $8 million in cash dividends.

The Chicago Lumbering Company was said to be the most efficient pine lumber company ever to operate in Michigan—a very broad statement indeed. I think it was true for the following reasons:

- The owners were experienced lumbermen who had made fortunes lumbering in New York and Pennsylvania before they came to Michigan.

- From 1871 until the 1890s, Manistique and Schoolcraft County were isolated in winter from the outside world except for weekly mail brought from Escanaba over the ice, usually by an Indian, to Fayette and from Fayette to Manistique by snowshoe courier. After the Detroit, Mackinac & Marquette Railroad, now the DSS&A (Duluth, South Shore & Atlantic Railway) was built, the mail was brought by stage down from Shingleton. In summer, access was by lake boats.

- George H. Orr, General Woods Superintendent, and M. H. Quick, Superintendent of the Mills, Yards and City Property, knew from personal experience the exact requirements of every man's job and how it should be done. There wasn't a trace of high hat and any employee could talk to the bosses anytime as man to man.

The result was a close-knit organization which operated in complete harmony for over forty years, just like a big family, with no change at all in management and almost no labor turnover.

Linderoth Studios, Manistique; W. S. Crowe collection

The seven Orr brothers, pioneers prominent in Manistique history for a half century. Standing, left to right: Fred, Walter, Burton, Erastus. Seated: Aaron, Abner, George H.

C. L. Company's "Mud Hen" paddle-wheeler used for rafting and booming logs on Indian Lake in the 1880s and 1890s.

W. S. Crowe collection

Many of the men worked for the company for their entire lives and became experts in their jobs, especially in the woods. In the rare cases when a lumberjack sought a job outside, he only needed to show that he had worked satisfactorily for the C. L. Company.

George "G. H." Orr was Woods Superintendent for the C. L. and W. L. companies for forty years, and President of the C. L. Co. The Orr Bros. partnership—Erastus, Burton and Walter, along with Ed Brown—built the Orr block and had a meat business, cattle ranch and slaughter house. Fred Orr was sheriff for four years,[7] a busy office in those days, and Walter was village president at a time when the caucus resembled a three-ring circus or battle royale. Abner Orr, barn boss, had charge of the C. L. Barn, the Indian Lake Farm and about three hundred heavy draft horses. There was no nepotism. If anything, the Orr brothers were tougher in business dealings among themselves than with others.

George H. Orr came to Manistique with his family on the schooner *Express* in 1872 and contracted to put in logs for Weston on Murphy Creek. He did so well and made so much money that Weston bought up his contract, gave him a five percent stock interest in the C. L. and W. L. companies and made him General Woods Superintendent. In later years he was made President of the C. L. Co., an office he held until his death in 1912. His five percent stock interest in the W. L. Co. made him a millionaire.

During the forty-two years the companies operated, their mills never lost one minute's time for log shortage. There were always logs—usually a reserve in the rivers that was sufficient to run the mills for a whole season.

The first Sunday I was in Manistique, John Woodruff and I hiked out to Indian Lake and then walked across on the big pine logs that filled more than half the lake (the company's old tug, as big as a house, was out beyond the logs)[8] to the present golf course—then part of the company's Indian Lake Farm—to see the horses. There were over 275 big heavy draft horses out for summer pasture, any one of them fit for a ribbon at a horse show. I had seen plenty of horses on the western cattle ranges, but nothing like these. As I walked among them, now and then one would raise its head, look mildly at us for a few seconds and then calmly resume grazing. They evidently decided we were harmless creatures of little importance.

Mr. Orr once told me that "some horses have more sense than some men," and I thought of a certain farmer on the river road who sometimes got well pickled when he came to town; later, friends would load him, in an unconscious stupor, into his wagon or sleigh and unhitch and start the horses. The horses never failed to take him home, where they would stand and wait patiently until relieved.

George Orr was a great lover of horses and a good judge of horseflesh. He had a fast-stepping little mare and a light cutter for his camp inspection trips, and foremen said they liked to watch him speeding along, horseshoes clip-clopping and sleigh bells jingling, over the wide icy log roads, smooth as a boulevard, winding in and out among the trees. He would inspect the horse barn first, and the surest way for a foreman or teamster to get in bad with him was to mistreat a horse.

John Creighton was a tote teamster for the C. L. Co. for nine years. He was an expert horseman in whom George Orr had a lot of confidence, and on one of his horse buying trips outside he took Creighton along. On his trips to and from camp Creighton would shoot deer. The camp bought all he shot on the way up to camp for $2 a head, and the deer he killed on the way back he sold in town at $3 a head. I was told he shot forty-nine deer one fall on these trips. His rule was, "one cartridge, one deer," and he never took a chance of wasting a shot or wounding a deer that was too far away for a sure shot. In those days there were no game laws, deer were very plentiful, the "sportsmen" hadn't yet arrived on the scene and lumbermen and the Indians killed only what they needed for meat. Many of the camps had a camp hunter to keep them supplied with meat. George Hovey told me that the hunter for the C. L. Company's camp on Doe Lake killed over seventy-five deer one winter without having to leave the camp. There are probably five times as many deer killed or wounded in the deer season today as there were then, and back then we never heard of men, cows and horses being shot and killed because someone thought what they saw was a deer.

M. H. Quick was a millwright and saw filer for Weston for years before he came to Manistique in 1872 to become General Superintendent of the Mills, Yards and City Property, continuing until the company sold out in 1913. He was also one of the officers of the company, in which he had a ten percent stock interest.

When two men can direct the operations of a big company for over forty years, with stockholders and directors in complete agreement and with practically no labor turnover, it is certain that both management and men had to be just about tops. Mr. Orr had twenty or twenty-five of the best camp foremen in the country, and Mr. Quick's mill and yard foremen and crews were organized to the last detail.

Further proof of the company's efficiency is the fact that they delivered millions of feet of clear white and red pine lumber on the boats at an overall cost of $9 to $12 per M, including stumpage. White pine stumpage was figured at $3 and the mill rate for custom sawing other companies' logs that happened to go through the mills was also $3. On those prices, the company made a profit.

CAMP TALES

Frank Cookson was one of George Orr's most efficient camp foremen and rivermen. The story is told that Cookson bet he could cut a log in the woods, carry it to the river and ride it across without getting wet, and when someone took him up on it, he cut a dry cedar about twenty-five feet long, carried it to the bank and not only rode it across the river, but lay down on his back and smoked a cigar while doing it. Some doubt that, but I knew Cookson pretty well and I believe it. I know he could easily have carried a dead cedar log big enough to float him, and there were dozens of river drivers in the C. L. crews who could ride and birl logs and perform stunts better than many professionals I have seen. The Knuth boys were especially good.

Cookson and I once went up to inspect and check George Roberts' operations. Roberts had finished a large cedar contract on the Upper West Branch of the Manistique River. We went to Shingleton on the Haywire[9] and from there walked on snowshoes over seven miles, arriving at his No. 1 camp about dark. All the camp watchman had to offer were some mouldy biscuits, cold soggy boiled potatoes that had began to spoil and fat salt pork—"sow belly"—without a streak of lean in it. We made tea and fried some of the potatoes which helped a little, but Cookson told me afterward it was the worst meal he had ever eaten. As for me, however, after a seven-mile hike on snowshoes I had the appetite of a goat. We spent the next day inspecting Roberts' cuttings, which covered six sections, and then hiked back to Shingleton the same night.

For me, that was a journey into the never-never, a giant's fairyland. The deep snow clung in the most unbelievable shapes to the stumps and tangled slashings; odd and misshapen tree trunks leaned at all angles; blackened ghostlike stubs from old forest fires and living green cedars all mingled in the utmost confusion to create a perfect foundation for the snow to form a fantastic wonderland.

There were caverns, tunnels, bridges, steeples, castles, draperies and lace curtains, and shapes no description would fit, and everywhere the gigantic toadstools—enormous snow caps on the stumps. One on an eighteen-inch stem measured eight feet across. The absolute silence and dead stillness was without a sign of life and these grotesque and monstrous shapes

State Archives of Michigan

A "stump mushroom" of snow stands tall in the deep woods of Ontonagon County.

State Archives of Michigan

A winter trip in the Carp River valley, Marquette County, 1890s.

in the dim twilight were almost frightening, but from another angle the unearthly beauty of the scene in the ghostly moonlight, with the Northern Lights flickering overhead, surpassed all imagination. Even Cookson, a practical man familiar with woods scenes, remarked about it.

That night I dreamed I was a pygmy lost in a wilderness of giant toadstools peopled with tall black monsters reaching for me with long arms. The dinner probably had something to do with my dreams.

I was also semi-aware of an unfamiliar rustling sound which later turned out to be just a bunch of lively young teenager bedbugs out making a night of it under my pillow.

My ideal old-time lumberjack was Lyman Timmerman, a lean and muscular six-foot, two-inch New York Stater who had worked for Weston at Gang Mills (Painted Post), New York, and on the Allegheny River drives. He came to Manistique in the late 1870s and left about 1904 for California where he bought a ten-acre prune and apricot ranch in the Santa Clara Valley near Mountain View. I visited him there twice and, although the Santa Clara Valley especially at blossom time seemed like a corner of paradise, he said, "Yes, Bill, it's nice, but I miss the woods. I miss the wind in the pine tops. I miss the early morning breakfast gong, the cry of 'timber,' the snow, the frosty creaking of the log sleighs, the sprinkler and the ice roads and the big horses straining in their harness to start the load, and G. H. (George Orr) with his fast little mare and cutter on a round of inspection. I miss the dinner

State Archives of Michigan

Logging camp bunkhouse, Delta County. Sunday was the loggers' only time for laundry and leisure.

Lumber camp crew in Schoolcraft County. Only a few backwoods logging camps had female cooks.

Schoolcraft County Historical Society

horn echoing through the timber, and the evenings in the bunkhouse around the camp stove, the shoe pacs and the steaming German Sox hung up to dry. I even miss the straw bunks of the earlier days, and I'd give all this to bring those days back. That was a man's country."

Timmerman eventually sold the ranch, which had cost him $3,000, for about $25,000 and moved up into the Washington timber country. He seldom wrote a letter and each time I saw him he handed me some money to pay his Masonic dues back in Manistique.

Lyme Timmerman could swing his heavy double-bitted axe over his head and split a wooden match with it nine times out of ten. To most of us an axe is just an axe, but not to a real lumberjack. His axe had to fit him and be just-so as to balance, weight, hang, length and thickness of handle, and of razorlike sharpness, usually with one blade ground thinner than the other. The double-bitted axe is not only a mechanical tool but a very dangerous weapon in the hands of a greenhorn, although during the seven years I kept the company's books not a single accident was reported from the woods. It speaks volumes for the amazing skill of those old-timers when eight hundred to a thousand men could work in logging camps and on river drives with double-bitted axes, crosscut saws, cant hooks, peavies and pike poles year after year without accident.

Edgar C. Brown came to Manistique in the early 1870s from Sharon, Pennsylvania. He was yard foreman in charge of the shipping (boat loading) crews until the company ceased operations. His job in the winter was to scale all logs put in by jobbers, or purchased from other owners. I made many trips to the woods with him, and on those occasions would tally the logs for him. He told me interesting stories of his experiences in the earlier days. In the winter of 1872 he was given some papers to record at Munising (then the county seat of Schoolcraft County).[10] He made the snowshoe trip through snow six feet deep, accompanied by an Indian. There were no trails or landmarks, his only guide being a compass. Using the compass he would point out a tree to the Indian, and when they reached that point the Indian would say, "make him talk again." It took them five days to make the trip—two days going, one day in Munising and two days returning—sleeping out both nights under balsam trees, the limbs of which they would bend down to make a sort of tent. On the return trip they were joined by landlooker John M.

Longyear[11, 12] of Marquette, then on his way to the Fox River section. Brown told me about brushing out Cedar Street for surveyors and about a Jim Ward trapping a wolf on the site where the Presbyterian Church now stands. At another time when he was running some lines near Indian Lake, he said he stopped for a drink at a spring, and when he turned around an Indian stood there who said, "White man come back seven summers." The Indians had a legend that anyone drinking from this spring would come back in seven years, and naturally most of them do, which proves the legend to be correct.[13]

George Hovey was an outstanding camp foreman and river boss for the C. L. Co. for many years. I once spent three days with him on a camp and woods inspection trip. Someone told me he couldn't swim, and when I asked him how on earth he could boss drives and ride logs on the river without learning to swim, he simply said, "I never had to swim." [14]

NORWAY LOGS WENT DEADHEAD

A C. L. Co. camp foreman told me this story: One of the company camps above Seney had finished its cut, and in the mopping-up operations, the foreman cut some "small" Norway pine trees from a knoll near the camp.

George Orr came along and, noticing these logs, said, "What in tunket did you cut those for?" George Orr was the only one of the old C. L. crowd whom I ever heard swear, and he summed his profanity all up in one word, "tunket." I have searched the dictionaries, and made diligent inquiry, but have never been able to find a definition of the word "tunket." It must have had a very comprehensive meaning, because George Orr was continually using it on every occasion when things didn't go to suit him.[15]

When Mr. Orr asked the foreman why he cut those small Norway logs, the foreman said, "Well, they were handy to get and we were cleaning up, so I thought we would take them along." Mr. Orr said, "You leave them right there," and they were left there.

In those days of winter logging and summer river driving and sawing, it was very necessary to guard against uncertain weather and unforeseen contingencies by having a large reserve of logs in the river ahead of the mills at all times.

After the C. L. Co. built the Manistique and Northwestern Railroad from Manistique to Shingleton in 1896, they pulled the logs out of the upper Indian River at Steuben, hauled them about twenty miles and unloaded them into the Manistique River at the "C. L. Dump" about four miles up the river from Manistique.

But before the railroad was built they drove the logs all the way down the Indian and across Indian Lake. Julius Phillion told me that he was in charge of the Steuben "pull-up" the first year of its operation and that they loaded out 48 million feet to the rail line and it took another year just to clear the floating logs in the upper Indian above Steuben.

Phillion estimated that the "reserve" of logs in the Manistique and Indian Rivers and their tributaries, and in Indian Lake and the chain of lakes at Uno, was not less than 250 million feet, or enough to keep the three mills busy for over two years.

The distance by river from above Seney to Manistique is hardly less than 125 miles, and it might be from three to five years before the logs cut at the camp reached Manistique, during which time a large proportion of the small Norway would go "dead head." Norway is heavier than white pine, and small Norway logs are heavier in proportion than large ones.

Then, too, it would be human nature for the river drivers, used to handling the large white pine logs which would float like a cork and were easy riding, not to pay too much attention to a comparatively few small heavy Norway logs. Mr. Orr knew a good share of them would go "dead head" and that quite a few of them would be lost in the drive.

On top of all that in those days it is doubtful if the company was making any money on Norway pine. They certainly were not on small Norway logs.

No one knows how many million feet of "dead head" logs lie on the bottom of the Manistique and Indian Rivers, Stutts Creek, the West Branch and other streams and in Indian Lake, and the other lakes tributary to these rivers, but there is no question about who owns them. They are now owned by Harvey Saunders of Germfask.

One year when Julius Phillion had charge of unloading the cars at the "Dump," George Orr asked him to go down and see how many logs they had in the Jamestown lakes. These were sizable bodies of water formed as an offshoot of the Manistique River, located east of the river just opposite from where the Indian River empties into the Manistique from the west. Phillion had no trouble walking all over the "lakes" on the logs which then filled them, many of which had been there for years. They were a small part of the immense reserve of logs which the company kept in the rivers and lakes at all times.[16]

FEELINGS OF FREEDOM

I have often wondered where the artist who painted the mural in the lobby of Manistique's new Post Office (for which I am told the government paid $900) got his inspiration.[17] He certainly never saw anything like that in the woods. The lumberjacks never used pole axes, which were something for the tote teamsters to carry on their wagons and sleighs, or for women or chore boys to split kindling with, and the trees the government lumberjacks are working on may have grown in a cow pasture but never in the woods.

Those old-time expert lumberjacks formed a sort of undefined aristocracy and to belong one had to be able to give and take it. They were completely loyal to the Chicago Company, and there was a spirit of rivalry extending from the camp boss even down to the cookee and chore boy as to which camp could put in the most logs or haul the biggest load.[18] Hardship meant little or nothing. The lumberjacks would have scorned featherbedding. Hard work was the rule for employers and employees alike. Every man in those days was strictly on his own in a really free world. But nobody starved, and when unforeseen calamity fell, the company, fellow employees and the neighbors always helped out. There was no feeling of insecurity, and we faced the future with boundless optimism and faith in ourselves to handle any situation. Every American boy was a mechanic and familiar with firearms, and the idea that any foreign nation would attack us on our own ground was laughable. On the other hand, we didn't carry a chip on our shoulder and dare anyone to knock it off.

Lynn McGlothlin Emerick

Mural of a logging scene painted on a wall of the Manistique Post office, and dedicated in 1940. In Crowe's opinion, this painting did not accurately represent real woods logging operations in Upper Peninsula forests.

Workmen of today enjoy many privileges we didn't have. There are modern inventions undreamed of then, and cost of living measured by man-hours, the only true measure, instead of dollars, is much lower now than then.

We didn't have commentators to think for us. We had to do our own thinking, and anyone who thinks thinking isn't hard work should try it sometime.

We had no government bureaus to plan our lives. We had to do our own planning. On the other hand, we didn't feel called upon to plan the lives of half a billion Europeans and a billion Asians, or to finance the development of countries in every corner of the earth.

I soon sensed that the lumberjacks regarded me for a long time as just a kid, a mere white-collar pencil-pusher office clerk, although they never said or did anything to hurt my feelings. I simply wasn't important enough for them to notice one way or the other. For my part, I idolized those he-men and looked up to them with a sort of awe. When one of them condescended to talk to me and call me "Billy" I was extremely flattered and pleased and felt that perhaps I, too, belonged in a way.

As time recedes into the past, men, events and movements little-appreciated at the time loom up in perspective, just as at close range the foothills obscure the peaks, but from a distance—let's say from the observation platform of the eastbound Rocky Mountain Limited—the foothills fade out and the peaks rise majestically from the plain until, one hundred miles out, only the highest snow-covered peaks are still visible above the flat horizon.

So it is with human life. As time recedes and the perspective clears, many little mannequins on the world stage fade out and disappear, and the truly great loom up as the mountain peaks of history.

And, looking backward, certain characteristics of those old "horse-and-buggy" days, little noticed at the time, now loom up in perspective.

It was an age of direct action.

A jobber came in to see George Orr about logging a tract of timber on C. L. Co. lands. He objected to the terms, but when Mr. Orr reviewed his logging plans and showed him how he could improve them, he closed the deal. Mr. Orr then made a pencil memo to Mr. Mersereau, who dictated a letter which Orr and the jobber initialed, and a copy was handed to me as bookkeeper. The whole deal, paper work and all, didn't take thirty minutes, and it was typical. If the jobber had met Mr. Orr on the street they would have made the deal right there. Direct action, right off the bat.

To see company officials, you didn't have to give your name and state your business to a front office girl, make an appointment and perhaps be told they were busy or in conference and to come back later. They didn't seclude themselves in private offices. Their high roll-top desks were right out in the open accessible to the public, and there was no formality. You simply walked in, stated your business and got your answer then and there. You wouldn't be put off or told, "We'll think it over and let you know. Come back next week." There was an absolute minimum of paperwork and everybody had more time. To close their offices for the day they just pulled down the tops of their roll-top desks and went home.

They could do business that way because they were owners and not managers for other owners, and they knew the company's affairs so well that they didn't have to "think it over." They had already thought it over, carried the answers in their heads and were always ready for business.

George Orr carried a pocket plat book and I think knew every forty in Schoolcraft County. When the annual tax list (a big one in those days) was published in the *Pioneer-Tribune*, Mr. Mersereau would call off the descriptions and Mr. Orr told him at once whether to "let it go" or "buy it in." Mr. Mersereau pencil-checked the ones to buy in, handed the list to Miss Riggs, the stenographer, who listed them in a letter that went to the country treasurer with a check, and that was that.

In the little one-room school house (No. 2, Adams Township, Butler County, Pa.) from which I graduated at the ripe age of fourteen, there were just two mottos on the bare walls. One, "Knowledge is Power"; the other, "Time is Money." The organization of the Chicago Lumbering Company of Michigan exemplified these mottos to the nth degree.

One sensed that they were men of calm judgment and great reserve power and resourcefulness, who were never in a hurry and had complete confidence in their ability to handle any emergency that might arise.

In the seven years I was in the C. L. office, I never saw Mr. Quick, Mr. Orr, Mr. Mersereau or any other company official lose his temper, get into a heated argument with an employee or high-hat or do anything to humiliate any employee. The humblest employee could meet and talk with any of them on the level, and that attitude was perfectly natural without a trace of affectation. Nor did I ever hear one of them tell an off-color story or utter an oath. Mr. Mersereau taught a Sunday school class in the Presbyterian Church and told us one day that folks who used a lot of profanity in everyday talk, whose every other word is an oath, aren't usually bad people, but simply ignorant folks who lack the vocabulary to express themselves otherwise. I still think of that when something tempts me to swear.

BARGE AND MILL

Abijah Weston was a tall, lean, powerful man with a fringe of white whiskers who reminded me of pictures of Horace Greeley. He commuted two or three times each summer between Tonawanda, New York, and Manistique on the *Buell*, the flagship of his Tonawanda Barge Line fleet of twelve ships. The

C. L. Company owned a three-fourths' interest in the Tonawanda Barge Line (TBL).

It took the *Buell* and her barges several days to load, and Mr. Weston would spend the time inspecting the mills and yards, with an occasional trip into the woods with George Orr. Quite often, he would bring back a pine stick into the directors' room and, absorbed in his thoughts, sit and whittle by the hour, leaving the floor covered with several inches of shavings when he left. He was a silent man, but very polite, and we in the office stood rather in awe of him. He had other timber holdings aside from his controlling interest in the Chicago and Weston lumber companies and the Weston Furnace Company, then located at the present site of the Manistique city garage.[19] He sold his Weston Furnace Company, which was succeeded by the Burrell Chemical Company and the Lake Superior Iron and Chemical Company, afterwards reorganized as the Charcoal Iron Company of America.

A. Weston & Son also owned one of the largest yards in North Tonawanda, then the leading lumber distributing port in the country, and was heavily invested in power developments at Niagara Falls. I happened to be in Tonawanda when the Niagara Falls Power Co. was building a power house, and Mr. DeGraff, manager of A. Weston & Son, gave me a card to the supervisor. I found him down at the bottom of a two hundred-foot penstock where they were installing a turbine wheel. It gave me an eerie feeling to hear the rumbling of the mighty falls only a short distance away.

Mr. Orr and Mr. Quick, when not out on the job somewhere, would sit in the directors' room, where they each had a desk, and visit and sometimes, apparently daydreaming, would just sit and relax, saying nothing. Although they never seemed to be in a hurry, they could move fast enough in an emergency but seldom had to because they were always one jump ahead of it. They thoroughly believed in the old adage, "A stitch in time saves nine," and they instilled this belief into the employees.

Inside of an hour after the mills shut down in the fall, Mr. Quick would have a crew sweeping out, cleaning up, checking, repairing and greasing all the machinery, tools, belts, saws, tram cars, locomotives and so on, and in a short time the mills were in perfect order ready to start at a moment's notice, although it might be five months before they went into action again. They also took off all the slashboards and opened the gates in every dam from the headwaters of the rivers to Lake Michigan, to drain the rivers, lakes and swamps, and provide a natural reservoir for the annual spring freshets.

When Mr. Orr planned his woods operations to supply three big mills with logs, he didn't plan for one year, but for the next year and the year after next, because with the main river full almost to the High Rollaways and the Indian River and half of Indian Lake full of logs, there was always a supply ahead for this year. In the forty years Orr supervised the woods operations, not one of the mills ever lost a minute's time for lack of logs.

A tote teamster told us this story: Two young Swede lumberjacks were hiking to camp with their turkeys on their backs. Said Ole, "Ain't a gon to be much snow dis winter." Said Nels, "How ya no?" "Yorge Orr, he tole me so." He said someone repeated this story to Mr. Orr as he was figuring some camp reports. Without lifting his eyes or cracking a smile, he said, "I never told him any such thing," and kept on with his figuring.

"Brigham" Young was a colorful old-time lumberjack, one of the best in the Chicago Company's outfit, who always blew his stake and had a good time after being in the woods all winter. In those days of winter logging, spring and summer river driving and summer sawing, with skidding and hauling done by heavy draft horses, and sometimes even with ox teams,[20] on log sleighs and sometimes big wheels—no chain saws, Caterpillars, logging trucks or paved roads in those days—men went to the woods in the fall and stayed until the camps broke up in the spring.

If all the lumberjacks had been like Brig, there would have been some foundation for the stories that pictured them all as a "red-eyed, rough and tumble, hell-raising fighting gang" who thought only of whiskey and women, although Brig himself was not a quarrelsome man. He was just a good fellow and when he got to the mellow stage, he would say, "Well, I must go down and talk things over with J. D.," meaning J. D. Mersereau, Secretary-Treasurer of the W. L. and C. L. companies.

Like many outdoor men, Brig had a well-balanced perspective of life as it is, and a shrewd native wit, and his talks with J. D., who also had a fine sense of humor and gift of speech, were a high spot in entertainment for the office force. Brigham would lean over the low counter by J. D.'s desk, and J. D. would carry on a running commentary with him for an hour or more, while keeping right on with his work.

Brigham took his responsibilities for the company very seriously and J. D. had the tact and finesse to make him feel right at home, as if the company were just waiting to hear from him. He couldn't have carried it off any better if he had been planning the company's future with Mr. Weston himself.

And when Brigham talked himself out, he would go back to the St. James Hotel on the west side where many of the lumberjacks stayed, and tell the boys all about it—how he and J. D. were going to run things from now on.

Another office event was my doing. It happened when I was filling in as time boy and Miss Moore, a timid, rather pretty little thing, was assisting Miss Riggs during summer vacation. It was the age when practical jokes were in vogue. I had a rubber worm, a coil about two inches long, which, when released, would uncoil and look and act just like a live snake. The dignified directors were holding their annual meeting in the adjoining room with the door wide open. Coming in from my rounds I said to Miss Moore, "Hold out your hand and I'll give you something nice." She did, and when I dropped this crawling snake in it she let out a yell that could have been heard in the next county, and kept on screaming. I thought of an errand up town, jumped on my bicycle and pedaled away from there as fast as I could. Fully expecting to be fired, I didn't show up at the office until the directors had all gone; but when Miss Riggs told me they didn't even look up from their deliberations or take any notice of the commotion, I breathed easier. I fancied I saw a twinkle in J. D.'s eyes when he greeted us the next day with his usual "Good morning, gentlemen," but I never did get back in Miss Moore's good graces. She was off me for keeps.

I mention these incidents simply to show that the C. L. and W. L. companies' officials were not easily disturbed or caught off balance.

CHAPTER TWO

At the Sawmills: Cribs, Yards and Office

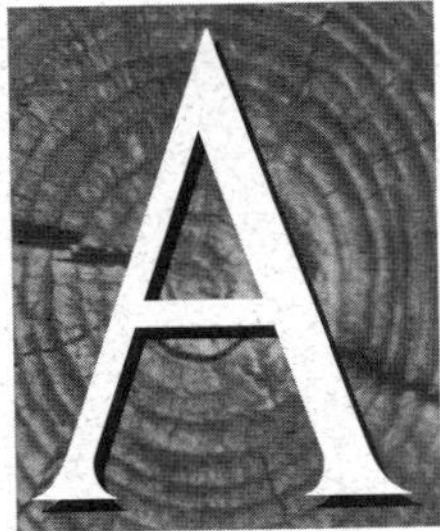

A charming little miss came up to me in the lobby after the high school concert the other night and asked if I could tell her what a "crib" was. It flashed through my mind that she was just ribbing me (with three teenage granddaughters I am quite used to that), and it was on the tip of my tongue to tell her it was something to keep babies in and ask her how she ever managed to get out. Then I sensed from the look in her eyes that she was really "honest and truly, cross my heart, hope to die" asking for information.

So I told her that the cribs in the river were big, square affairs made of sixteen- or maybe twenty-foot logs, bolted together at the corners and filled with heavy rocks. The cribs rested on and were anchored to the river bottom and extended about six feet above the surface of the water at normal level; their purpose was to back up the booms.

Then I told her that a boom was a string of long logs or tree trunks, chained together end to end, or perhaps square timbers bolted together, which floated on the surface of the water, some of them being movable, so as to control the logs in the river and divert them into the proper channel to the mill which was to cut them.

A line of these cribs and connecting booms extended for miles up the center of the Main River (the loggers never called it the Manistique). About one-half mile above the Soo Line bridge was the main boom where cedar logs, tie cuts and shingle bolts were sorted out for the White Marble Lime Company's shingle mill, then located on the east bank of the river just north of the Soo Line.

About a mile up the track from the present Intake Park was the dump, where the C. L. Co. dumped logs brought down by rail from up by Steuben [21] into the Main River and just above that was a pull-up, where Hall & Buell pulled logs out of the river to ship by rail to their big steam mill at South Manistique; still farther up the river was the Sorting Gap where logs were sorted and diverted to the proper channels.

Then just below the Soo Line bridge, and above the upper dam, a crew of river men sorted out the logs for W. L. Mill No. 2 and shunted the rest through the chute in the dam to go on down the river to be again sorted and divided between C. L. Mill and W. L. Mill No. 1.

Schoolcraft County Historical Society

Hall & Buell Company's sorting gap on the Manistique River.

So it will be seen that the entire river for seven or eight miles above Manistique was like a railroad yard where cars are classified and switched to certain tracks just as logs were sorted and switched to the proper river channels. And it was a very lively scene with river crews guiding the constant stream of floating logs, using their long wooden pike poles with sharp point and hook ends to push or pull the logs. These river men were expert log riders. I saw one cross the river on a cedar tie cut without getting wet.

A large floating boom, three feet wide and two feet deep, formed of clear twelve-inch by twelve-inch white pine sticks forty-eight feet long, bolted together, extended from the C. L. Mill pond on the east side to W. L. Mill No. 1 pond on the west side, just above the lower dam, in the form of a "V" pointing upstream. Its purpose was to divert the logs to either mill, and to keep them from going over the lower dam and out into Lake Michigan. About four hundred feet long, anchored to and backed by five or six heavy stone cribs, it floated with its top perhaps six inches above the water and was sometimes used by foot passengers as a shortcut across the river. The lumber in that boom would be worth about $10,000 on today's market.[22]

Soon after I started work in the C. L. office I had to fill in as time boy for several months, to collect the scales, time books and sales slips from the mills, stores, docks and other offices, and get them to the office by 10 A.M. I found that I could save some time by riding across the river on the boom. Mr. Quick saw me one morning and said, "Will, it makes me uneasy to see you riding your bicycle on the boom." I didn't take that seriously until a week later when he saw me again, and said, "Will, I think I mentioned that it made me uneasy

A crew sorts logs at one of the pull-ups near the mouth of the Manistique River.

F.A. Baker photo; Jack Orr materials, by permission of E. Marion Orr

Jack Orr materials, by permission of E. Marion Orr

Sorting logs near the C. L. Company's mill, 1892. The timber, after coming down the streams, entered this sorting gap to be directed to the proper mill. The old railroad bridge is in the middle distance. Other buildings are fish sheds and warehouses.

to see you riding the boom. I hope I won't have to mention it the third time." I said, "I'm sorry, Mr. Quick, and I won't do it again."

It required no special ability to ride a bicycle on a walk three feet wide, even if it was floating on the water, but I didn't argue the point with Mr. Quick. I realized what a mill foreman told me, that Mr. Quick was a very mild-mannered, soft-spoken man who never barked out an order, but that his wishes were orders as far as the men were concerned. By starting at 6:30 instead of 7 A.M. I could still get the reports in by 10, after which I had office work to do until closing time at 8 P.M. I thought I was lucky to have a job with such a fine company, and this feeling was shared by ninety-nine percent of the employees.

HIGH-WATER DAYS

The highest water ever seen in the Manistique River was in 1914. It was almost up to the girders of the Soo Line bridge, and I had Ab Gage and a crew of thirty river men piling sand bags on the levees above for days. We weren't afraid if it got past the bridge, as both dams below were wide open with all slash-boards off, and although they were completely submerged with a solid sheet of water three feet deep pouring over the tops, we knew they would hold.

I said to Mr. Quick, "You must have had pretty good engineers when you built these dams," and he said, "Our mill and river crews built them with the help of millwrights and a surveyor who knew how to run levels. There was a twenty-five-foot rapids in the river, and we had no data or records to go on, so we studied the shores and the timber along the banks and could tell the highest point the rivers had ever reached, and built them with a good margin above that. We made sure they were big and strong enough to hold."

In 1917 an "expert" outside engineer from New York City planned the present flume leading to the papermill and the highway siphon bridge to replace the old iron truss red bridge and also a concrete bridge for the Ann Arbor RR to replace its old iron truss bridge just above the highway

W. S. Crowe collection

The Goodwillie Box Factory was heavily damaged during the 1920 flood in Manistique, with the loss of 125,000 board feet of lumber.

bridge. Looking at this new railroad bridge with its deep girders and only two open spans, and also at the changes being made at the upper dam, some of us went and told him that he wasn't leaving enough channel for the river. Looking down his nose at us, he said, "This flume will carry all the water that will ever come down that river."

But just three years later, although the water wasn't as high as it was in 1914, it broke through the new works and did over $500,000 damage to west side property, wrecking the Goodwillie Bros. box factory, doing a lot of damage to the paper mill, washing out streets and railroad tracks and ruining a lot of homes with the water and mud into their second stories.

For a week or two, the only communication between the east and west sides of the city was on foot over the Soo Line bridge. After this happened the bottleneck under the Ann Arbor RR bridge was enlarged.

Prior to 1920, Manistique never suffered from high water damage, and it was also about the only Michigan lumber town which was never seriously threatened by or suffered from a forest fire.

And speaking of high water damage, a disastrous flood on Indian Lake in 1943 did a lot of damage to resort properties and cottages, a contributing cause of the damage being the costly mistake made some twenty-five years ago by the county engineer when he diverted the Indian River from its natural channel at the outlet, cut a new channel through a sand bar and built a new bridge without the capacity to take care of emergency floods.

In June 1945, after considerable discussion and a petition signed by ninety-five percent of the property owners on the lake, an official level was established by court order, and it is the responsibility of the County Board of Supervisors under the law to maintain the level.

THE GANG AND THE BAND

Hall & Buell's big double-band steam mill at South Manistique was running full on smooth logs one sunny afternoon in the 1890s when a distinguished group of men walked through on a tour of inspection. At their head was Abijah Weston, and with him Alanson (A. J.) Fox, N. P. and W. E. Wheeler, George H. Orr, M. H. Quick and J. D. Mersereau. These men owned ninety-eight percent of the stock of the Chicago and Weston lumber companies, then operating three large water power gang mills in Manistique.

Cutting band saw teeth in filing room, Hawley Sawmill on the Dead River, Marquette County, 1890s.

Marquette County Historical Society

The purpose of their visit was to determine whether to change over and install band saws at Manistique to replace the slower and more wasteful gang saws.

A band saw is a thin steel belt, fourteen inches wide when new, with teeth on one edge; it runs over two large wheels about eight feet in diameter, one above the other, at a speed of 10,000 feet (or about two miles) per minute. The logs are fed to it by a carriage shuttled back and forth so fast that the two men riding the carriage have to be real Rough Riders to stay on.

A gang saw is a rack of large thick saws set in a vertical frame that oscillates up and down, cutting only on the down stroke, the saws being spaced at varying distances to cut lumber of the thickness desired.

The gang saw cuts the entire log, or "cant," in one operation, whereas the band saw cuts only one board at a time, but very fast.

Hall & Buell logs came in by rail from a pull-up on Indian Lake and another one on the Manistique River just above the C. L. dump. They were unloaded into a log pond in which river men on floating booms sorted them into pockets. In the vernacular of the mill hands, the logs were Little Rough (small knotty logs from the tree tops); Big Rough (larger and not so knotty from the lower limbs) and Smooth (the big clear logs from the bole of the tree below the limbs). From the log pond an endless chain carried the logs up the jack ladder onto a raised deck in the mill, from which they could be rolled either way onto the carriages which fed the band saws. One side was the Big Side; the other, the Little Side.

In the Hall & Buell mill the logs were usually slabbed on four sides with the band saws and the square cant then went to a large gang saw in the center of the mill. This big, steam-powered gang saw moved up and down so fast that it was almost a blur

Marquette County Historical Society, from *St. Nicholas Magazine*, 1897

A gang saw in operation. The early saw mills used water power to run gang and circular saws.

compared with the slower water power gangs in the C. L. and W. L. mills.

The minority stockholders were sold on the merits of the band saw, but Abijah Weston, who owned fifty-one percent of the stock, couldn't be budged. He was sold on the merits of his perfect gang-sawed lumber—in fact, advertised it—and in all fairness to him it must be said that the only lumber to equal it is the four-square lumber turned out in the precision mills of the giant Weyerhauser Co. on the Pacific coast. I am told that the only remark Mr. Weston made after this visit was that he wasn't going "to have his lumber cut by lightning."

About this time Mr. A. J. Fox, Vice President and the most influential stockholder next to Weston, came out in the office where I was working on the books and said, "Will, I wish you would figure out how much money we would have saved on last year's cut if we had had band saws instead of gangs. I'd like the figures for the directors' meeting tomorrow afternoon." I said, "All right, Mr. Fox," but I thought, "Oh, boy, this is it." I worked late that night, and was down at daylight the next morning, but when I handed Mr. Fox my figures about 1 P.M. and he said, "That's about what I figured," I felt well rewarded. As I remember it, the figure was around $93,000 and would have been more except that a large percentage of that year's cut was two-inch and thicker stock. In other words, the thick gang saws were cutting about $93,000 worth more of clear pine lumber into sawdust than the band saws cutting a kerf only half as wide would have done.

The heavy gang saws in the three mills in Manistique cut perfect lumber and lots of it, but in cutting one hundred million feet of one-inch by twelve-inch clear white pine boards,[23] they chewed another twenty-five million feet into sawdust. Band saws would have saved at least ten million feet of that, worth $150,000 at the average mill run price of $15 per thousand board feet prevailing in the 1890s. A local dealer told me that he would have to get from $400 to $450 per thousand for clear one-by-twelve white pine, at which price a season's cut on one hundred million feet would be worth about $40 million, and the gang saw waste would be worth $4 million.

We criticize the old-time lumbermen for wasting our resources, but wait a minute. How about the settlers in Pennsylvania and Ohio who cut the finest white oak, black walnut, elm and bird's-eye maple trees, rolled the logs into piles and then burned them to clear the land? Just one single black walnut tree four feet in diameter (they ranged from

Jack Orr materials, by permission of E. Marion Orr

Chicago Lumbering Company sawmill, 1908. Logs, separated by size and type, were floated into the chain conveyers and taken up into the saws.

Chicago Lumbering Company office.

W. S. Crowe collection

three to eight feet) would be worth over $4,000 today.

Now, Michigan lumbermen were doing exactly the same thing as the Pennsylvania and Ohio farmers. They were cutting trees and clearing land to build houses and make farms. The only difference was that the Pennsylvania and Ohio farmers built houses and made farms on the same fertile soil from which they cut the trees, while the Michigan lumbermen cut trees from sandy pine barrens to build houses and make farms out in Illinois and Iowa.

In those days good farm land was at a premium and timber was cheap because there was so much of it, working under the old law of supply and demand.

But who are we to pretend holy horror at the waste of a few millions by the old-timers, who at least had the excuse of what looked like inexhaustible natural resources to work with, when we, with the end of our natural resources in plain sight, are wasting billions to their millions and are even giving away our income and mortgaging our children's future at a rate unparalleled in all history?

LUMBER'S PAPER WORK

To keep the books and do the paper work for the Chicago and Weston lumber companies—with a payroll of 1,200 to 1,500 men, cutting around one hundred million feet yearly and operating twenty-seven retail departments dealing in every item (except drugs, whiskey and fresh meat) required by Schoolcraft County's seven thousand people isolated from the outside world four months in the year—we had an office force of five people: A head bookkeeper, payroll bookkeeper, assistant bookkeeper, stenographer (Miss Riggs) and a time boy, whose job was to collect the sales slips, cash, coupons and time books from the stores and foremen. His route covered the east and west sides of the river from the main boom (the present Wyman Nursery site on the river) to the docks, and he had to have them in the office by 10 A.M. The general office was a room about thirty feet square. A flat top counter separated working space and public lobby. Doors, always wide open, led from the lobby and the working space into the directors' room where Mr. Orr, Woods Superintendent, and Mr. Quick, Superintendent of the Mills and Yards, each had a roll-top desk. Mr. J. D. Mersereau, Secretary-Treasurer, had his roll-top desk in the working space with one end next to the counter, directly accessible at all times to the public and the office force.

Jack Orr materials, by permission of E. Marion Orr

Whereas, The CHICAGO LUMBERING COMPANY and the WESTON LUMBER COMPANY are about to employ me to work for them in their lumbering operations and about their railroad, and in various ways connected with their lumbering operations; now, therefore, I agree that, in consideration of such employment and as one of the conditions thereof, I do hereby waive the provisions of the act of the Legislature of the State of Michigan, passed in 1885, making ten hours a day's labor, and agree that a day's labor shall be the same hours as has heretofore been their custom according to the direction of said Lumber Companies, or the person who has charge of the work in which I am engaged, be the same more or less than ten hours, and this agreement is to continue in force so long as I am in the employ of either of said Lumber Companies.

Date .... 188 ....

Witness ....

A reproduction of a waiver signed by lumberjacks before being hired by the Chicago or Weston lumber companies. This precondition to employment was in common use around the time the original was signed in September 1885.

The assistant bookkeeper and time boy worked off the main office in a side room that also housed the telephone switchboard with its thirty drops, most of them to the company's own departments; "tending central" was part of their job.

The Manistique Telephone Company, wholly owned by the Chicago Lumbering Company, was just a debit account on their ledger; the total capital investment was less than $1,500. It kept growing and the C. L. Co. incorporated it about 1896-97 with capital stock of $5,000, paying themselves a $3,500 dividend. My brother Dean purchased $500 of the stock at par, managed the company for ten months, then sold his stock for $1,400 and went west to engage in the telephone business on a larger scale in Oregon. The company continued to grow, paying big dividends and financing itself almost entirely from earnings until 1928, when Mr. O. G. Quick sold it to outside brokers for $45,000. Promoters capitalized it for several times that amount, sold preferred stock to Manistique residents and kept the common. The company is now a unit in the General Telephone Co. system.

We also kept the books of the Manistique & Northwestern Railway Company, which the Chicago Company built in 1896.

The only mechanical device in the office was a Remington invisible typewriter with an inch-wide purple copying ribbon, and a copy press with which all outgoing letters were copied in an indexed letter book. We had no carbon copies, card indexes, vertical files, calculators, bookkeeping or adding machines or loose-leaf records of any kind. To qualify as a bookkeeper, one had to be a good and rapid writer, quick and accurate with figures, and well grounded in spelling, grammar and geography. All bookkeeping was with pen and ink in bound books. An old-time bookkeeper's adding machine was in his head, and the speed with which some of them could foot a column of figures was almost uncanny. I was far from the top, but even today, out of practice, I can set down the total of a five-man bowling score at a glance. You don't add it. You picture it, beginning at the hundreds column. A sheepherder counting a band of sheep going through a gate pictures them in groups of five or ten.

The six hundred-page C. L. and W. L. flat-opening, bound ledgers, of the finest heavy Byron Weston linen ledger paper, with a heavy canvas cover, weighed fifteen pounds each and would stand a lifetime of handling. A leatherbound indexed signature book also lay on the counter and every man who worked for the companies had his signature in that book.[24] These bound books of the C. L. and W. L. companies from 1871 to 1914 were a pretty good history of Manistique, and it is unfortunate that they were eventually sold as waste paper to the paper mill.

Bookkeeping is very simple. Its essence, whether for United States Steel or the corner grocery, is a journal, a daily record of the business done, from which it is posted into classified accounts in a ledger. All other books and reports, no matter what business it is, are auxiliary to these two.

We made great use of columnar forms. The C. L. payroll, written up each month, consisted of flat sheets, twenty-four by thirty-eight inches, with horizontal lines for employees' names and vertical columns for entry in the general ledger at the end of the month.

One advantage we had with these bound records partly offset the advantage of modern office appliances. We didn't lose time, or our tempers, hunting for lost, misplaced or misfiled papers. We knew where everything was. The only loose things were incoming letters filed in Amberg flat files. Sales slips, camp orders and so on were filed away in neatly labeled shoeboxes.

Another advantage was that there were no income, sales, inheritance, social security or unemployment taxes, or price fixing codes. We had no government or state reports whatever to make until 1896 when the C. L. Co. built the M&NW Railway, after which I made a monthly report to the Interstate Commerce Commission, but it was far from the detailed reports required today.

As a mass economy with modern machinery, we today are of course much more efficient in production, but it is a question open to debate whether the personal efficiency of the individual is as high as a general rule as it was in those days; whether we have the initiative and resourcefulness of individuals who had to plan largely on their own.

I believe in invention and progress. Stability is stagnation and death. I always liked my work, but I never did a tap of work in my life without trying to figure out some way to shorten it or eliminate it entirely.

Tom McCullough was platform boss at the C. L. Mill. When the lumber came from the mill proper, it was carried sidewise by an endless chain over a wide, long-roofed sorting table, where it was graded, sorted and then loaded onto tram cars on the platform, a floored space large enough to hold the night's run of seventy-five to one hundred thousand board feet to be hauled out to the yards in the morning. Mr. Quick came along one day on his early morning round of inspection, and McCullough pointed to the platform covered with tram cars loaded with clear two-by-twenty-four white pine planks, and some even three-by-thirty-six, and said, "There's something, Mr. Quick, that'll put a plug hat on you." Stovepipe hats were quite the thing and a symbol of wealth and distinction in those days. M. H. Quick smiled but said nothing. This was tops in grade and commanded a very high price, maybe as much as $40 or even $50 per thousand. I was told that a lot of it went to England for patterns. Great Britain was then the dominant industrial nation, with her big iron and steel plants in Birmingham, Manchester and on the banks of the Clyde, although the United States was catching up fast.

CHAPTER THREE

The Waterfront

From the summer of 1863, when the first lumber schooner dropped anchor off the mouth of the Manistique River—about the time Confederate forces under Robert E. Lee reached their high-water mark of the Civil War at the battle of Gettysburg, through the 1880s and Gay Nineties—the waterfront was a scene of great activity in Manistique, as it was in other lumbering towns on the Great Lakes.

For about thirty years all freight and passenger traffic in and out was by water. Boats were constantly coming and going. The Goodrich Line of Chicago ran a boat once or twice a week as far as Manistique, touching at all intermediate Wisconsin, Green Bay and Lake Michigan ports, including South Manistique and Thompson on the way.[25]

The Harts Line ships, *Fanny*, *Eugene* and *C. W. Moore*, served all Green Bay and Lake Michigan ports as far as Petoskey and Cheboygan, and there was a Harts boat in nearly every day. A fast steamer, the *Hunter*, made daily round trips from Manistique to St. Ignace and Mackinac Island. Weston's TBL fleet, *F. R. Buell*, *A. Weston*, *Allegheny* and *Canisteo*, each with two or three barges, were busy all summer carrying lumber to Tonawanda and bringing back supplies.[26] They all carried passengers and the *Buell* was popular with Manistique people revisiting their York State

A Harts Line steamer in Manistique Harbor. The Harts Line served all lake ports on the north shore of Lake Michigan from Green Bay, Wisconsin, to Cheboygan, Michigan, in the 1880s and 1890s.

W. S. Crowe collection

Superior View Studio, Marquette

A crew loads lumber from the Delta Lumber Company sawmill onto a schooner at the Thompson harbor, Schoolcraft County.

homes. Hines boats came in often and independent lumber schooners were continually coming and going.[27] I once counted twenty-nine lake boats alongside eighteen lumber schooners tied up in the harbor and the slips. The waterfronts were busy in all lake ports. The *Lotus* shuttled back and forth daily between Escanaba and Gladstone, and a daily boat ran from Escanaba to Fayette.

Quite often a Goodrich or Harts boat would run moonlight excursions, very popular with the young folks, out into the lake for two or three hours, taking either the Woodsmen of the World band or the City band with it. The Harts boats also ran excursions to Mackinac Island and Petoskey now and then.

All this was new and very fascinating to me, and I spent as much spare time as possible rowing

State Archives of Michigan

The sailing schooner *Quick Step* carried lumber from Ontonagon and other Lake Superior ports to Thunder Bay, Ontario, and down the Great Lakes to New York mills.

W. S. Crowe collection

Indians sailed fleets of these "Mackinaw" sailboats from the Beaver Islands to Manistique, where the families sold their handicraft. These two-masted open boats were fast sea boats.

a boat or sculling a yawl around the harbor, up the slips or out to the old breakwater. One evening, sitting on the breakwater, I saw three schooners put out under full sail in a race for the Chicago market, and the setting sun on their sails was a beautiful sight. The *Hunter* coming in from the straits, rolling like a barrel and throwing spray high over her deck, added to the picture.

It is not generally known that, in a gale, a sailing ship rolls less than a steamer of the same size.

MACKINAW BOATS, FISH TUGS AND GOSPEL SHIPS

It was a picturesque sight in the spring when the Indians sailed a fleet of their Mackinaw boats over from the Beavers[28] bringing whole families, from papoose to grandmother, with moccasins, baskets and other Indian handicraft for sale. These two-masted open Mackinaw boats,

Early postcard; W. S. Crowe collection

Indian basket sellers on the path to the Big Spring (Kitch-iti-ki-pi) in the early 1900s.

W. S. Crowe collection

A fishing boat in Manistique Harbor. The man in the center of the photograph shows off a catch of suckers to those on board.

eighteen to twenty-four feet overall length and about eight feet abeam, were very fast and good sea boats, and the Indians were excellent sailors. The women would sit in the boats, surrounded with their well-made handicraft, or wander through the town, many with papooses on their backs, carrying baskets. They never asked you to buy, but they always sold out quickly.

A. Booth & Co., leading fish dealers in the U.S., had a fleet in Manistique, their *Oval Agitator* almost as large as a lake steamer.[29]

A colorful character seen often in the harbor was Captain Bundy with his gospel ship, the *Glad Tidings*, a converted fish tug. He was an itinerant waterfront missionary, familiar in all northern lake ports. I often heard him talk to the sailors. He could talk their language and was quite successful with them.

After navigation closed and the mills shut down for the winter, Mr. Quick started his crews cleaning up and overhauling the mills, and eight or nine hundred lumberjacks packed their turkeys and trekked to the lumber camps, most of them not to be seen again until the spring breakup. For the next four or five months, Manistique was completely isolated from the rest of the world, except for a weekly mail carried across the ice, usually by an Indian, from Escanaba to Fayette, and from Fayette to Manistique by snowshoe courier. Ed Brown told of some of his trips and about stopping overnight at the halfway house this side of Garden. After the Canadian Pacific Railway built the Soo Line in 1889, we had a daily mail.

The C. L. Company was more than just a lumber company. It was the commissary for a community, including the county, of nearly seven thousand people shut off from the rest of the world all winter. Every fall the company shipped vast quantities of goods and supplies of every description and nature into the Manistique harbor from Green Bay and Chicago, via Harts and

Goodrich boats, and up from York State and Detroit, via the TBL: Oats, salt pork, flour, oleo, potatoes, onions, candy, German Sox, shoe pacs, Mackinaws, underwear, bedding, hardware, furniture, leather, Peerless tobacco, condensed milk, coffee, tea, sugar and spices, clothing and dry goods, Percheron horses, ladies' lingerie and baled hay—every conceivable article of use. There was a company doctor to bring you into the world and a company hearse to take you out of it and up on the hill.

LOADING THE LUMBER SHIPS

The Steamer *Pahlow* of the Edward Hines Lumber Co.'s fleet steamed into Manistique harbor on a July morning in the late 1890s. The captain wanted to get loaded and out the same day to avoid losing two days on account of the holiday, and he asked Ed Brown, boss of the C. L. Co. shipping crews, to help him out. He was asking quite a favor because everybody was getting ready for the Glorious Fourth, the greatest national holiday of those times.

Nevertheless, Brown told him to get the inspectors while he rounded up two crews, or forty men. Benjamin Gero, manager of the inspection firm of Martin, Silliman and Gero, asked W. J. Shinar, the only inspector available, if he could handle it. Shinar agreed. Carl Eckstrom was tally boy and Charley Robinson checked the piles for the C. L. Co. The loading started shortly after 8 A.M. and by 5 P.M. the boat was loaded and ready to go. It was a good job well done in a hurry.

The loading crews were divided into three gangs to start—one below to stow the lumber away, with a second on deck to take it from the third gang on the piles and pass it down until the hold was full. Then the deck load was piled and secured.

The tramways from the mills to the yards and down between the slips were about fourteen feet

A good catch: Sturgeon at A. Booth & Sons Fishery warehouse, Manistique. Corwin Adkins and Seymour Graham hold the fish stake. Both Booth & Sons of Chicago and John Coffey, who came from Fayette in 1880, were leading fish dealers with extensive fleets operating from the port of Manistique.

W. S. Crowe collection

high.[30] The standard lumber pile was sixteen feet front and about twenty feet high and contained between 20,000 and 25,000 board feet; because the *Pahlow* carried about 480,000 feet, it required about twenty piles to load her. The crews worked a "four-pile front" so that the boat had to be shifted several times during the loading. The inspector had to be almost on the run to inspect and measure four piles of lumber as fast as two crews could load it, and the lumber shovers didn't keep out of his way—he had to keep out of their way.

A simple calculation will show that to load the *Pahlow* in eight hours meant loading nearly a thousand feet per minute, or four pile courses. Obviously, the inspector didn't have time to measure each board. It wasn't necessary. He would count the boards in a course, and if there were, say, fifteen one-inch by twelve-inch by sixteen-foot boards, that would be 240 board feet, or 960 feet in four courses; but even so, he had to move fast to travel seventy-five or eighty feet from one end to the other of a four-pile front, jump the gaps between the piles, measure and inspect the lumber, chalk-marking any culls or "scoots" and making due allowance for the twelve-foot and fourteen-foot lengths usually piled in with sixteen-foot boards. In one hand he carried a flexible lumber scaler's rule of hickory, about forty-eight inches long with a handle at one end and a brass ferrule at the other. A board could be measured without bending down, or flipped over for inspection by a twist of the wrist. In the other hand, he carried a marking stick with a crayon clamped in the end to instantly mark with an X any culls or scoots found in a merchantable pile; these were then thrown off on the ground. The largest percentage of the mill cut in Manistique in those days was one-by-twelve white pine boards in twelve-, fourteen- and sixteen-foot lengths. The lumber was piled in courses with three "crossers" between each course. Pile bottoms were timbers, and the piles were roofed over by standing a plank on edge at the front, from which boards were laid overlapping and sloping to the back of the pile; to make it secure, crossers were laid on top and fastened to courses down in the pile with iron clamps.

When the yard crews finished a pile of lumber, the company inspector would estimate and mark the total feet and grade with black paint on a course down in the pile front, for example, 23480 WP 1x12 or, say, 22160 WP Culls 1x12. I am told that the difference between the company's estimate as marked on the piles as checked by Charley Robinson and the inspector's scale on the *Pahlow*'s load was less than a thousand feet.

Lumber inspection fees were twenty-five cents per M, the buyer and seller each paying half. The inspector's scales and grades were accepted as final, and settlements between buyer and seller were made on that basis.

The scene looking west from Pat Quinlan's three-story boarding house near the C. L. Mill was one of great activity, with boats loading, puffing tugs and dinky engines, creaking tram cars and screaming saws, gangs of leather-aproned lumber shovers, inspectors calling tally, lumber clapping on decks, Andrew Ekstrom's busy shipyard (where Sellman's fish tugs now tie up) and a smoky haze from the burners over all.

The pine has now been gone for many years; and the fleets of steam barges and picturesque schooners with their tall masts in every harbor and white sails dotting the blue waters of the Great Lakes are only a memory. The scene today, looking west from the same spot, is one of utmost desolation, a silent, dreary, grass-grown waste with rotting docks, a dead world without a sign of life—and that is the scene most of us envisioned for Manistique when we were singing the swan song, "When the Pine is Gone." And it is what might have been.

But facing east we now see that the sun rises over a modern city with up-to-date stores, fine homes, public buildings and well-lighted paved streets utterly unlike the uncouth lumber town of the 1890s.

We had heard something about a harmless crank somewhere down around Detroit who was wasting his time tinkering with a "horseless carriage," to the great amusement and wisecracks of the crowd. Little did we dream that the material future of Manistique and many another town in the sticks hung on the development of that funny horseless carriage, the forerunner of the modern automobile.

The development of a nation's economy and way of life is analogous to that of the automobile. First this part and then that. First good roads. Then tires which would stand up. Then springs which wouldn't be continually breaking. Then a self-starter. You couldn't run a modern car with the carburetor of that day. Every new gadget makes necessary an improvement to some other part of the machine. So it is with a nation's life. We have made astounding strides in our material world in the past fifty years, but our moral and spiritual development has lagged or even perhaps deteriorated, and we will now have to catch up with that end of it or we will be ditched and perhaps destroyed by a wonderful machine we have created but seem unable to control.

Superior View Studio, Marquette

World's Fair log load—Shipped from the Upper Peninsula to Chicago for the Michigan exhibit at the 1893 World's Fair.

CHAPTER FOUR

The River and the Woods

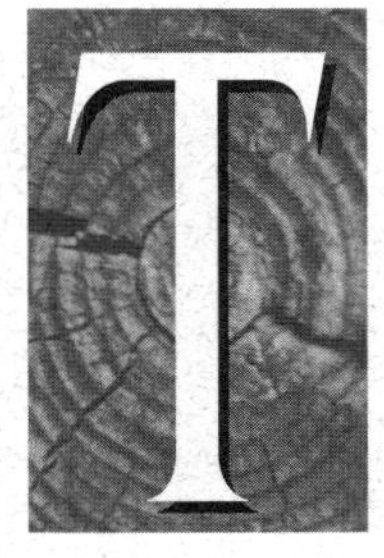

The big sawmills of the white pine era in Michigan were almost all located on deep water on Lake Michigan or Lake Huron, usually at the mouth of a stream.[31]

In the 1880s and 1890s, from 120 to 126 million feet—all pine—were floated down the Manistique River every year.

Ownership of the logs was divided among four or five companies. Most of the logs were sawed in the three Manistique mills, but many millions of board feet were handled at Hall & Buell's big steam double-band mill after being separated out at a sorting gap upstream from the Manistique mills and moved by railroad to South Manistique.

A large amount of timber was sawed for Alger, Smith & Company, which had a mill at Grand Marais and owned timber on lands north and south of the DSS&A Railroad. The Chicago Lumbering Company also owned land in that timber and often cut forties for Alger, Smith and floated the logs down the river to be custom-sawed in the mills at Manistique.[32]

The work of the river hogs, as the men engaged on the drives were nicknamed, was extremely hazardous. They had to see that the logs were kept continually on the move, and when the logs jammed and piled up in a huge mass, it was their job to break out the jam. They would go out on the jam with their peavies to release the key log. When this log was freed, the water and the logs roared down the stream, and it was a wild scramble to safety for the river driver. Once in awhile one of them would be crushed and lose his life in the jam, but this was rare, according to Frank Cookson, who said, "The Manistique is an ideal river for drives. It has a uniform depth all along the route, and is free of sand bars and rock. In all the years I spent on it, I remember only four men who lost their lives."

The principle of driving three million feet of big logs four, five and six feet in diameter down a small stream is exactly the same as swallowing a pill.

A lady once told me she had a terrible time swallowing pills. She said she put them as far back in her mouth as she could and had great difficulty getting them down.

I told her to hold the pill between the tip of her tongue and her front teeth, and then take two or three swallows and release the pill on the second or third swallow. She did so, and told me she had no further difficulty whatever.

State Archives of Michigan

Double-gated sluice dam on the Indian River, Schoolcraft County, 1888. The sluice, or splash, dam was opened when water was needed downstream to float the logs.

That is exactly the principle on which logs are driven down the river. There has to be water ahead of and behind the logs, which are released at the crest of the flood. If the logs were put in the river and then water turned on, they would create a jam and dam up the water and it would be impossible to get them afloat. But when the river drivers were ready to drive, water from the dams upstream would be released very early in the morning, probably five or six hours before the logs were to be put into the river. It had to be done at just the right time, and there had to be flood water behind the logs, to keep them afloat.

Perhaps the finest tract of big pine in the Upper Peninsula, located a hundred miles from Manistique by river, was about 25,000 acres owned by the C. L. Company and the Weston Lumber Company around Driggs Lake and the headwaters of the Driggs River. This was magnificent tract pine, with many four- and five-log trees, four to five feet and even more in diameter.

The Driggs was only one of the streams down which logs were driven to Manistique. There was also the West Branch, the Fox River, Stutts Creek and Indian River, and every stream had to be handled differently. The Driggs, the Indian and the Fox all had dams to hold the floods and save water. These would be closed at night, and opened for the drive for a certain length of time, say seven or eight hours, in the day.

The C. L. Company had a series of dams from the headwaters of the Indian to Manistique in the following order: Hartney, Harrigan, Little Tote Road, Big Tote Road, Doe Lake, Six-Mile, Ten-Mile, Steuben, Jackpine and the present M-94.[33] When the mills shut down in the fall, Mr. Quick opened all head gates, removed the slashboards and drained the whole river system, lakes and swamps as low as possible, thus forming a natural reservoir to retard the annual spring floods that were certain to come next year.

The actual mechanics of cutting those big trees at the headwaters of the Driggs and getting them down to Manistique involved four operations. A three-man crew consisting of a timber fitter and two expert axe and crosscut saw men would cut and limb the trees and then cut the trunk into sixteen-foot logs. The timber fitter's job was to size up the tree, decide where to make it fall and notch it accordingly, and he was very expert. In fact, the story is told that you could set a stake in the ground and the crew could fell a four- or five-foot tree so it would hit the stake and drive it into the ground.

The entire logging operation fell naturally into four parts:

1. *Timber Cutting:* Three-man crews cut the trees down, limbed them and sawed them into logs of sixteen feet, where possible.[34]

2. *Winter Skidding and Hauling:* Men and horses would skid or haul the logs out of the woods on narrow skidways to the wide icy logging roads, where immense loads of logs would be piled on sleighs and hauled by horse teams to be driven (floated) in the spring.

3. *Spring Driving:* Driving was perhaps the toughest job connected with the logging operations and required men of considerable technical knowledge and experience in floating and driving and riding on logs. It was the highest paid job in the entire operation.

4. *Summer Sawing:* After the ice had gone out and the streams were clear, the sawmills would start up as early in the spring as possible and saw all through the summer and fall as long as the weather permitted.

W. H. Osborn; W. S. Crowe collection

Sawyers fell a tree with a crosscut saw. The timber fitter has notched the tree on the side.

On the Driggs River there were two dams, the Ross Lake Dam and the Driggs River Dam. They would be closed in August to obtain all the water possible until the next driving time in the spring. The Ross Lake Dam flooded Ross Lake, Nevins Lake and some swamp land. This dam was used largely as a reservoir to hold the Driggs Lake Dam up to a certain level.

The first operation in driving a stream would be to see that the river was clear. A crew would use saws, axes and dynamite to clean out snags, stumps, trees and anything that had fallen into the river since the last drive.

W. S. Crowe collection

A filer touches up the teeth of a crosscut saw on a portable filing rack in the woods.

Schoolcraft County Historical Society

Log sleds and horse teams in deep snow of winter.

The stream was now cleaned out ready for driving, with the logs piled here and there at banking grounds along the river and on the banks of the lakes.

Frank Cookson had a drive camp about a mile and a half down the river from Driggs Lake, and logged and banked along both sides of the river and the south side of the lake, both above and below the dam. John Moran and crew had a drive camp one and a half miles south of the DSS&A Railroad and their logs were all banked in the river below Driggs Lake. William Bragg had a camp about eight miles below Moran's camp, and a man named Blake had the job of watching the logs that were put in by Ferguson Brothers of Munising. Cookson's job was to get the logs in the vicinity of Driggs Lake Dam and Moran's logs on the west side flowing down the river.

LEGACY OF A LOG

Let us for a moment trace the life history of one of those big logs. It came from a family group which were tiny seedlings when Columbus set out for the Americas. It was about 130 years old when the Pilgrims landed at Plymouth Rock. One hundred years after the birth of the new nation it still showed no signs of old age or decay. The group of trees stood near the shores of an inland lake, and the great trunks, eighty or ninety or even one hundred feet to the first limb, were like the columns of an immense cathedral.

It was a quiet scene, the ground carpeted with pine needles with no underbrush, and the moccasined feet of Indians passing on the trail now and then made no sound. Far above, the breeze sighed in the majestic pine tops—no sound like it in all the

W. H. Osborn; W. S. Crowe collection

A sprinkler used to ice the skidding roads. Note the four water streams from the back of the sprinkler. The wooden pins in the box on the rear of the sleigh were used to close the sprinkler holes. Icing of the roads was done at night, when the temperatures were below freezing.

Superior View Studio, Marquette

Lumberjacks on an Escanaba River log drive, circa 1890s. Each man had a peavey for moving and guiding the logs.

world.[35] The scene changed; many years later a troop of big rugged men with teams of great horses and sleighs came into the woods. The men wore brightly colored mackinaws, and carried heavy double-bitted axes and long crosscut saws. Then we heard the sound of many axes chopping and saws whining and much shouting and singing, and pretty soon the cry of "tim-b-e-r," and the echoing cry of "tim-b-e-r, tim-b-e-r" came through the woods all the long afternoon, and after each ringing call of "tim-b-e-r" a forest giant would come crashing down with an earth-shaking jar.

The largest tree ever cut by the C. L. Co. was on the Driggs River, northwest of Wolf Lake above Seney, in the winter of 1898-99.[36] The butt log, twelve feet long, measured seven feet two inches at the small end and by Scribner decimal rule would scale 4,310 feet. The second log, sixteen feet long, was five feet, two inches in diameter and scaled 2,890 feet. The clear white pine lumber in those two logs on today's market would be worth considerably over $2,000. It took the three-man gang (William Pine—appropriate name—timber fitter, and Louis Stribbage and Ed Heitzman, sawyers) over half a day to fall the tree.

They had trouble driving this Driggs River log, because it was too big to float clear. When the four or five men bringing up the rear got one end out in the stream, the current simply rolled it over to the other side to hang up again. Cookson came back to see what was the matter and told them to leave it there. He sent Harvey Saunders for a large auger and some dynamite and blew it into three pieces; Harvey rode these pieces down the river to catch up with the drive.[37]

On the Driggs River drive, when the drivers were ready to start, they would open the dam before daylight and let the water run for about an

W. S. Crowe collection

Banking ground along the Indian River at George Hovey's Camp 76—T44, R18, Section 29. This camp was near the Chain of Lakes in Schoolcraft County.

hour before sluicing any logs through; otherwise the logs would travel faster than the water and run high and dry because the water had to fill up each little bend or opening along the river. Then they would start to "break the rollways," meaning to put the logs afloat in the river. The drivers had teams with a line to pull the logs off the center as they were rolled in. The breaking line had a hook and another small line to pull it back. They would break out 100,000 feet of logs in about twenty minutes, and then four men would start down the river riding the logs, about ten miles by river to where Moran's crew would take over at his "beat." Beat meant the section of river allotted to each crew. Following these first four men, about every twenty minutes to a half hour, two more men would ride the logs. The object was to try to run the logs through below the DSS&A Railroad in one day; in that year they had very little trouble doing that. If there was trouble or a jam, they would work on that jam until they figured they had about the same amount of logs, say 100,000 feet, ahead of them as they had started with.

After all the logs were through the dam, they let the water run for an hour and a half, and four or five men, the last of the upper crew, would start down the river, leaving the dam operating one hour after the last man started down the river. The whole idea was to keep the logs moving.

RIVER DRIVING AND LOG JAMS

River driving was tough: Long hours, sometimes in waist-deep water, wrestling big logs off sand banks, wading swamps or walking narrow pole trails up in the air, cold meals maybe and perhaps sleeping wherever night caught up with them, fighting no-see-ums and mosquitos.[38] No wonder drivers became agile and expert log riders and got over twice what regular woods work paid. If a teamster got $30 a month, a river driver would get $2.50 per day.[39] His $2.50, by the way, would buy more than $15 today.[40]

F.A. Baker photo; Jack Orr materials

Log jam on the Indian River, circa 1890. When logs piled up in the rivers during the drive from the woods to the mills, river drivers walked the jams and tried to free the key logs with long pike poles or peavies.

State Archives of Michigan; L'Anse Township Board

Log pile on the Big Huron River, Arvon Township, Baraga County, about 1910. When the spring floods came, these logs were washed downstream to Lake Superior, where they were enclosed in a boom or raft and towed to the sawmills.

The hardest work I ever did was on a canoe trip down the Indian River with Wes Orr when we ran into a cedar jam of John Hruska's and had to carry our canoe and outfit across a half-mile neck of an old pole trail mostly up in the air over a swamp of black muck. It took us three trips and four hours.

Logs that lodged along the banks were put afloat by "rearing," which was done with a pole about sixty feet long with a rope on each end. This rear boom would be swung out to stop a number of logs and thus raise the water in case of a sand bar jam in a bend of the river. In moving a jam, the men knew they had a clear space ahead, because two men had been sent over that part of the river a day or two before. About ten men from a normal crew of thirty would start breaking back. The foreman stationed these men at strategic places about a tenth of a mile apart, and each man stayed at the lower end of his beat. As long as the logs kept coming there was nothing to worry about; but when the logs stopped, the foreman waited until a nice bunch of logs went by and rode them down to find out where the logs had stopped. Then he came back to report the main jam a certain distance below. The cry "jam below" would echo from man to man until it reached the men breaking, who would stop until the flood was over, then get on the logs and ride right in on this space. This would be repeated until the rear of the jam was reached. There was sometimes trouble with the banks washing in case of a jam-in. Stumps would roll in. One man would be delegated to carry dynamite, and his tools were a pike pole, a bag of dynamite and an axe. With the cry of "jam below," the dynamite man would eventually get to the place of the trouble.

Ed Cookson was the walking boss, or the superintendent, and had a driving team. He visited the drive every day, and stopped either at the Driggs Dam or at one of the camps along the way. If water was wanted anywhere down the river, he would notify the man watching the dam and tell him what time to open. You could figure that he wouldn't be more than a half hour out of the way on timing the water coming down.

Logs were held in booms at the mouth of the Driggs until the waters of the Manistique River had receded—the big river sometimes overflowed its banks, and to drive the logs down on a flood would have meant losing many of them. The distance to the junction of the Driggs and Manistique Rivers is approximately twenty-five miles, and it took between two and three weeks to complete the drive to the Manistique.

When not more than four or five miles of logs were left in the river, the dam was opened and

State Archives of Michigan

Wanigans (cook rafts) followed the log drives and provided hot meals to the crews when possible. Some rafts had room for the drivers to sleep aboard at night.

allowed to run night and day. Usually it took three or four days from the time the dam was opened until the last log was in the Manistique River. The drives on the Manistique River were usually followed in a Wanigan, or crudely constructed large raft with a tarpaper shack that had a stovepipe sticking out of the roof. This was the home of the crew of twenty-three or twenty-four men during the drive; here they ate and slept, unless they slept in tents along the way, which they often did.

The Driggs River was one of the best to drive in this area, because a flood of so many hours was used, and a man, if he started at daylight, or four o'clock in the morning, could usually be home or back in his camp by one or two o'clock in the afternoon. On the West Branch of the Manistique River, in the Stutts Creek and in other streams of that kind, they drove on natural water and drove from daylight to dark.[41] The Driggs almost always had the best rivermen, and a riverman in those days was paid better than double what a teamster or a cant-hook man was paid in the lumber camp. On the river they fed workers better, too. They used to feed the men ham and eggs and such, whereas in the old lumber camps it was sow belly, red horse or corned beef.

The story in the previous paragraphs describing the Driggs River drive was told to me by Harvey Cookson Saunders; born in Maine in 1878, he settled in Manistique in 1897 and probably had more experience driving logs on the streams and tributaries than anyone else.

Harvey Saunders was brought up by his mother's parents, and his five uncles were all woodsmen and river drivers in Maine. When still a very young boy, he was fitted with cork boots like a river driver and was learning the trade. He came to Manistique with his uncles (Ed Cookson, walking boss for the C. L. Co., and Frank Cookson, a lumber camp foreman at Camp 60,[42] about six miles northwest of Driggs Lake) and drove the Driggs River in the spring of 1898 and again in 1899 and 1900. Mr. Saunders said, "From 1898 until 1935, there were only three years that I was not on a river drive on some river here in the Manistique area. I had charge of drives on the Manistique River from 1907 to 1934, except in 1918 and 1919."[43]

Mr. Saunders says,[44] "As far as I know, I am the only river driving foreman left in Schoolcraft County who was on the river drives in the days when the big pine drives came down, and I would like to name some of the old Chicago Lumbering Company's river foremen and the men who had charge of building most of the dams on the Driggs and Indian Rivers for the C. L. Co. Ed Cookson, one of the foremen, also Frank Cookson, Hartwell

Plumber, 'Red' Jack Smith, Charles Bridges, John Hartney, William Bragg, George K. Moody, James Stewart, Alex Rowe, Paul Knuth and George Hovey, foreman for many years on the Indian River. There may be others that I do not know."

Harvey Saunders was employed by the Interior Department in 1938 in connection with the Seney Wildlife Refuge until his retirement in 1956.[45] The next year he was honored by the U.S. Department of the Interior with a Citation and Certificate of Honor for Commendable Service, a Medal and Lifetime Pass to the National Parks, in recognition of his distinguished employment record with the Service. Harvey Cookson Saunders died suddenly in Texas on January 5, 1967, in his eighty-ninth year.

In the *Detroit News* of Sunday, June 30, 1929, is a featured article entitled "The Last Big Log Drive, Manistique Again turns into a River of Logs as Timbers Floated Down Recalling the Good Old Days." It is written by William J. Duchaine. As noted in the article, Mr. Saunders was foreman of the drive. Mr. Saunders purchased a river improvement company and was the owner at that time, driving logs both by contract and on a salary. But that is a story he tells best.

SAUNDERS' STORY

I, Harvey Cookson Saunders, purchased the Consolidated River Improvement Company from the Stearns Coal and Lumber Company. The Consolidated Lumber Company had purchased the River Improvement Company when they bought the Chicago Lumbering Company in 1913. I took over the company in 1926 and operated it as such until 1934. The boom company's job was to boom up the river where there were flooded areas where the logs might get spread out of the channel of the river, and to maintain the bridges up as far as Germfask. We had to provide booms and place them on the bridges to drive the timber through the bridges. We did not take over the handling of logs until they arrived at the junction of the Indian and the Manistique Rivers. From there on we handled all of the timber and sorted it or separated it, each kind, and placed it in a boom (or pocket, as we called it) by itself, to be turned over to the places where the owners desired to take it out of the river. For that work we received a toll of so much per cord and so much per thousand board feet for the various kinds of timber. It all had to be sorted out, balsam from spruce to go to the paper mill, logs to the sawmill (the Consolidated sawmill or the Stack mill). Later they had to be sorted. Sometimes they asked for the pine to be sorted from the hemlock. This was done by the so-called Consolidated River Improvement Company or Boom Company.

When we were operating the boom company, we owned many stone piers in the river to divide the river up to separate the various kinds of timber, as a man would fence off his field to put his

W. S. Crowe collection

Harvey Cookson Saunders, early resident and longtime lumberman.

sheep in one pasture and his cattle in another. It had to be in such a manner that the timber could be put to the loading place where the owner wished to take it out of the river.

In the month of March we had to obtain from the various lumbering men that were banking timber on the Manistique River an estimate of how much timber they expected to drive that year. That had to be reported to the St. Mary's Ship Canal Board. They set the price that we could charge the timber owners for sorting and running their timber through the booms in the improved part of the river. We had to pay a privilege tax, of course, and an incorporation tax, and we had to pay one percent to the Secretary of State on all tolls collected. I would estimate that at the time we were handling the timber on the river, we had about five miles of boom, or as a farmer would call it, five miles of fence, to fence the river off with.

We also owned a tugboat and we had two rowboats or river boats about twenty-five feet long. In making our report to the St. Mary's Ship Canal Board we had to include the value of our equipment that we were using to handle the timber. On that, partly, they based their price that we could charge for running timber through the booms.

I acted as foreman for the Consolidated Lumber Company on their drives and their sorting of the timber, starting in 1920 up to 1926. At that time, the Stearns Coal and Lumber Company took over and I purchased the River Improvement Company. It was a stock company composed of one hundred shares. One share was owned by W. B. Thomas, my secretary; one share was owned by F. M. Cookson so he could be on the Board of Directors. We operated the River Improvement Company for a long time, and it was a very smooth running company.

In purchasing the River Improvement Company from the Stearns Company, I also wanted the sunken and submerged logs, or deadheads, as some people call them. I knew I had at least five miles of boom in deadheads, some of it pine trees that were put into the river back in the Hall & Buell days, way before 1900; and they were thirteen inches at the top and sixty feet long. It made perfect timber. At the time I took them out of the river they were floating one-fourth or one-fifth out of the water. I knew that if someone else went on to reclaim those deadheads, it would be a constant fight for me to keep them from cutting up my own booms, scattered along the river for some distance; so I didn't want the River Improvement Company if I couldn't have the submerged logs, too. I never figured on making a fortune from them; my reason in purchasing them was to keep people from interfering with my own booms. They gave me a bill of sale of all the stranded and submerged logs with the marks that were used by the Weston Lumber Company and the Chicago Lumbering Company. The C. L. Co. had taken over what logs were left in the river by the Manistique Lumber Company. All marks used by the Consolidated while operating are registered and on file with the county clerk at the office up here.

Some people said that Saunders has the river rights on the river, but since the Consolidated River Improvement Company has been closed out, the only rights I have on the river are the bill of sale and the logs that have those various marks. I think a year and a half ago, I sold that bill of sale to Leonard Shay, of Germfask. That makes him the owner of the stranded and submerged logs that bear that string of marks. Those logs are marked on the end with a hammer and the mark is registered in the same manner that a stockman has his cattle or horses branded that run on the range. Anywhere Mr. Shay can find a log with the mark of one of those hammers, he can claim it, even if somebody picked up the old hammer and used it later.

I think that gives you a fairly good insight on the River Improvement Company.

We did have letters from the Secretary of State before I closed out this River Improvement Company stating that there were no other river improvement companies operating in the state of Michigan, whereas in the years past, any river of any size, where there was more than one lumbering man logging on it, had a river improvement company. It was necessary to have such a company because they were all fighting over the driving.

I don't know the date that the Manistique River Improvement Company was first started, but it dates way back. They spent thousands of dollars in building the piers and putting in the booms to hold and control the logs at the lower end of the river. When the Chicago Lumbering Company was operating, they had at times as much as one or two years' run left in the river over the winter. The main part of those were at what is known as the Stoney Cut, or the mouth of the Sturgeon Hole Creek; when the mills at South Manistique were running, that was where they sorted their logs from the logs belonging to the Chicago Lumbering Company. Their logs were loaded out of the Manistique River at what was known as the dump. It's the first place that the present "Haywire," or Manistique & Lake Superior Railroad, goes to the river north after it leaves Manistique going toward Shingleton. Just in the upper end of that high land, there was a mill. A pull-up was built in there whereby the mill at South Manistique and the mill in Thompson loaded their logs out of the river at the North Shore Gap or the North Shore Pier. It must be about two or three miles up river from where they had their pull-up or loading place.

The Chicago Lumbering Company logs came down on the east side of the river and the logs going to South Manistique and Thompson came down the west side of the river. There is a set of piers about every two hundred feet all the way from the present dump to the Sturgeon Hole Slough. At one time, they were in good repair.

And that was the way they handled the timber that came down the river.

After the South Manistique and the Thompson Lumber Companies went out of business, the Chicago Lumbering Company held the greatest part of their logs in the main boom that is right near the upper end of the present Wyman's Nursery. I think those old piers, if they haven't torn them down, are there yet. They were a good set of piers. There was a great deal of money spent in this river improvement company in the old days because those piers had to be built in the winter on the ice—then they were put in, large end at the bottom, and they put logs at the bottom and made a floor in it and built it up like a cobhouse camp. They cut a hole in the ice and started filling the structure with stone until the pier rested on the bottom. The timbers they used on the bottom on those piers kept getting narrower as they came to the top. Some of those old piers that are under water are still standing and solid yet. I think that a short time ago the sportsmen had quite a time getting them out of the lower part of the river here, from the Soo Line bridge up, for the boat races.

S. W. Peacock, Photographer; W. S. Crowe collection

A mounted Raymond log loader in operation.

A team of oxen skids logs from a cutting site to a deck along the river, where the logs will be piled for spring river driving.

W. H. Osborn; W. S. Crowe collection

State Archives of Michigan

Interior of a logging camp store, often also referred to as a Wanigan. The cost of goods purchased by the loggers from the camp store was deducted from their season's pay.

William Gibson, camp cook, and Ed Deloria, cookee, at a Schoolcraft County logging camp with six-foot dinner horn.

W. S. Crowe collection

W. H. Osborn; W. S. Crowe collection

A crew loads logs on a hauling sleigh, using cant hooks and chains.

Winter woods crew loads large logs in T45N, Range 17W, Section 20, near Stutts Creek, Schoolcraft County.

W. S. Crowe collection

S. W. Peacock, Photographer; W. S. Crowe collection

A steam-driven hauler with train of logs leaving the woods.

Harvey Saunders' logging camp, 1915-1917, in T45, R13W, Section 14, southwest of Germfask.

Schoolcraft County Historical Society

A big load of cedar posts from the Ed Cookson camp near Shingleton.

Superior View Studio, Marquette

CHAPTER FIVE

Kerosene Days

Wasn't life terribly lonesome in Manistique, shut off for four or five months every winter with only a weekly mail, no radios, movies or autos? Far from it. In fact, we thought we were living at the apex or peak of civilization, to which all history had been mounting. Hadn't we invented the mowing machine, the harvester, sewing machine, the steam engine, telegraph, repeater rifle, steamships and railroads. What else was there? I suppose the ancient Greeks and Romans felt the same way—that they were the climax of civilization with nothing much beyond.

Everybody was busy. Work hours were longer, but we didn't work at high pressure and most workers seemed to take pride in their craftsmanship. My own office hours were from 8 A.M. until 8 P.M. As head bookkeeper I got $83.33 per month and never felt overworked or underpaid. Good room and board at Mrs. Fuller's cost me four dollars per week. Has the cost of living gone up? Not at all. On the contrary, the cost of living today is less than half what it was in the 1880s and 1890s. But that is another story.

A bright young miss (I don't know who she was and I don't know where she came from, nor did I ask) stopped me and said, "Mr. Crowe, I like to read about the lumberjacks, but whatever did the rest of the people do, especially the young folks? It must have been terribly dull and lonesome in those old horse-and-buggy days, with no autos, movies, radios, airplanes, electricity or even refrigerators and with long hours, low pay and no vacations. You must have been bored to death. I don't see how young people stood it. I haven't any 'old folks' and the books don't tell much. I wish you would tell me something about them."

Well, to start with, the folks of that day were not so dead on their feet as the pictures might lead you to suppose. They managed to keep the race from dying out, long dresses, bustles and hoop skirts notwithstanding; the boys with their derby hats, stand-up collars and sporty vests were not so slow. The Gay Nineties were not misnamed. It was an era of large families, and time never hung heavy on our hands. We couldn't miss the modern gadgets now in everyday use any more than this generation misses the gadgets yet to be invented that will be in common use a hundred years from now.

There were plenty of parties, picnics, oyster suppers, church and lodge socials, masquerades, ice carnivals, amateur theatricals, Lyceum courses, Chautauquas, study classes, debates, parlor games, buggy rides and sleigh rides. The modern automobile is a thousand years behind Old

Dobbin when you're taking your girl out for a ride. You were perfectly safe driving with one hand, or you could even loop the reins over the dashboard and have both hands free. Old Dobbin was very understanding and would carry on himself. Contrast that with driving a modern auto at sixty miles an hour, nerves tense, eyes glued to the road, both hands in a vise-like grip on the wheel. At the end of a wild car ride, a young man asked his girl, "Well, glad you're here?" "Glad?" she said, "I'm amazed."

I wish I had words to convey the feeling of boundless enthusiasm, optimism and confidence in the future that pervaded America in the 1880s and 1890s. Curiously enough, England was America's bugaboo in those days. The eagle really screamed on the Fourth of July, and the orator who couldn't give the British Lion's tail a fancy little extra twist wasn't up to snuff. I don't think the English took it very seriously. Every small boy knew that one of Washington's blue coats could lick four British red coats any day, and, as for the grown-ups, we felt, "We licked them twice and could do it again," if we had to. Decoration Day[46] was a big day, with two bands, a parade and exercises. There were around eighty GAR veterans living here then and you couldn't have kept one of them out of the parade with a double-barreled shotgun.

It was the age of family life, fraternal orders, religious revivals, the bicycle craze and the Gibson Girl. The Gibson Girl, with her puffed sleeves, shirtwaist, sailor hat and long skirts, was a fetching attraction. She had brains enough to leave something to the imagination, and the young men of her day were not lacking in imagination. Families had their three meals at home and gathered around in the evenings for games, story telling or perhaps helping the youngsters with their school lessons. Housewives usually had a hired girl, prided themselves on their cooking and were continually exchanging recipes.

We were no more interested in continental Europe's continual revolutions and counter-revolutions than if they were occurring on another planet. We had eight flourishing lodges: the Masons, Knights of Pythias, Knights of Columbus, Maccabees, Odd Fellows, Woodsmen of the World, Foresters and the Norden—and the regular monthly meetings were attended by almost the entire membership.

We didn't have the hucklebuck or Spike Jones and had to do the best we could with Gilbert and Sullivan, John Philip Sousa's band and the waltz.

The churches were practically filled both morning and evening. Brought up in the Presbyterian tradition, I attended services there, including J. D. Mersereau's Sunday School class, Christian Endeavor and the evening sermon. But about seven months after I arrived in Manistique, John Woodruff introduced me to a young lady named Bertha Orr at a Saturday night Baptist skating party. After that I found Rev. Rooney's sermons at the Baptist Church very inspiring.

Ekstrom Brothers operated a large covered ice rink just behind the Lundstrom garage. It was the mecca for all young folks and crowded every night. Admission was ten cents, and fifteen cents twice a week on band night. The brass band played the "Skaters' Waltz" for the Grand March, and there were frequent races, figure skating contests, ice carnivals and masquerades. I made up as Kaiser Wilhelm for one masquerade, and the young lady with me was Mother Hubbard—quite a pair. After the Soo Line Railroad came, Jim Troyer, "champion backward skater of the world"

This 1902 photo, probably taken from the mast of a schooner in the Manistique harbor, shows drying fish nets and the boardwalk from harbor to Ossa Hotel. The men are Richard Hughson and Ray Stone.

From Jack Orr materials, by permission of E. Marion Orr

from the Soo, came over and easily defeated our best, with the exception of Peter Dube. Dube lived here and won a lot of races before moving to Escanaba. For his age he is without doubt the champion skater of the United States, if not of the world, and he defeated Troyer without much difficulty. Trot Lockwood, also from the Soo, gave exhibitions of fancy figure skating, but in my opinion Chris Drevdahl was as good as any of them.

Three livery stables, Fred Orr's on Wolf Street, Fran Vasseau's on Walnut and Harry Tucker's where the Oak Theater now stands, specialized in fine outfits, and it was the thing for a young man to take his steady girl sleigh riding, followed by an oyster supper in Ben Pollock's Oyster Parlor on the second floor of a building on Oak Street. Now and then fifteen or twenty young people would get together for a sleigh ride to Thompson, ending with a barn dance, which was about the last word in entertainment.

We had several newspapers in the 1880s and 1890s: the *Semi-weekly Pioneer*, published by Major Clarke, the *Tribune* by George E. Holbein and the *Härold*, a Scandinavian paper published by Nettie Steffenson and Carl Thorberg.[47]

Cows enjoyed complete freedom of the city. Many families owned cows, and every yard was enclosed with a board fence to keep them out. I once saw a cow leisurely window shop all the way from Putnam's Corner to where the Ford garage now stands. *Harper's Weekly*, which then occupied about the same place in American literature that the *Saturday Evening Post* does now, published a humorous article about the cows in Manistique, and soon thereafter our City Council passed a Cow Ordinance.[48]

AN UNBOUNDED FAITH

I've gotten carried away but, my dear young lady, it would take several books and a week's time to answer all your questions. Instead, I'll take you on an imaginary trip around Manistique in 1893," I said. And this is what I told her:

Schoolcraft County Historical Society

Cedar Street, Manistique, about 1884. Note the wooden sidewalks and fences. Orr Bros. Meat Market was managed by Burton Orr, one of the seven pioneering Orr brothers and Crowe's future father-in-law.

We are approaching the town on a large lake steamer. The city's two-story frame company houses, far apart on large lots, are spread out like wooden blocks on a board. There is not a tree in town because 'Bijah Weston, fearing forest fires, had them all cut down. Dominating the scene are three big sawmills running night and day. A hundred million feet of white and Norway pine lumber in high piles fill the entire yard; boats come and go; the tall-masted schooners are loading in the slips and harbor; the air is hazy with smoke from the open burners, picturesque lime kilns and charcoal iron furnace; and the scent of pine woods, fresh-cut pine lumber and cedar shingles, spiced with wood smoke from the open burners and the lime kiln, pervades the atmosphere.[49]

It is a busy scene we look at. Let's listen, too. We can hear the creaking wheels of tram cars, the whistling and puffing of the dinky locomotives hauling and bunting cars of lumber, lath, pickets, shingles and slabs; the echo of boards dropping on lumber piles, scows and boat decks; the incessant clip-clop of horses' feet on stone pavements; the puffing of harbor tugs and clanking of the big dredge which keeps one tug and a fleet of scows busy hauling sawdust out to dump in Lake Michigan; through it all, a deep undertone of

Linderoth Studios; W. S. Crowe collection

Chicago Lumbering Company's general store, later the Park Hotel.

sound—the Song of the Saws, big saws and little saws. They are singing a swan song, "When the Pine is Gone."

Now we are landing. If you have any cigarettes about you, get rid of them. "Coffin nails" are taboo, even for boys, and a man caught smoking one would be laughed out of town by "he-men" and lumberjacks who chew tobacco and smoke big black cigars. Everything is new—no vacant houses, abandoned plants or decayed docks—and there is an up-and-coming feeling in the air.

Let us start. First off, I can't take you around in that costume. We'll go into this big C. L. Store where they carry everything needed by a community of 7,000 people, from lumberjacks' Peerless to ladies' lingerie, perfume, baby powder and safety pins; the saleslady will fit you with a Gibson Girl outfit. And observe carefully—ladies never show any part of their limbs but to daintily lift your skirt and expose a well-turned ankle when crossing a muddy street is permissible. That and an air of general helplessness will make you very appealing to the boys.

Oh-oh! Get out of the way! Here comes a runaway, another thing that keeps life from becoming dull. A most spectacular runaway was a team that galloped down Houghton Avenue over three mill platforms and connecting tramways—wooden trestles high in the air with no side railing, only a little wider than the wagon the horses were pulling—across the river, down an incline from the C. L. Mill platform, then up Main Street to a dead stop in front of the Presbyterian Church.

Why did the horses stop in front of the church? Easy. They got religion. It is up-grade from the mill to the church, and when the going got too tough, they got religion. Just like people do when the going gets tough. They get religion. You may scoff and say horses can't think, but they do. Mr. Orr told me that some horses had more sense than some men. I have a broken collarbone from being bucked off a mean cayuse when I was only fourteen, although the horse afterward forsook its wicked ways and became as gentle as a lamb. There is a Horse Heaven, too. It is in Township 7 North, Ranges 23 and 24 East, Willamette Meridian in the State of Washington. If you don't believe it, look in Rand McNally's Atlas of the World, 1892 edition.[50]

An early bicycle shelter. Bicycling was a very popular sport in the early 1900s.

W. S. Crowe collection

W. S. Crowe collection

Manistique Bicycle Club, 1895. Will Crowe is standing third from left, back row.

The fact that the team stopped in front of the Presbyterian Church is pure coincidence and not to be taken as a reflection on churches of other denominations.

THE BICYCLE CRAZE

Next I'll take you around some of the mills, on a tandem bicycle ride to Indian Lake with a picnic lunch or for a walk through Indian Town or up to the furnace to see them cast, and maybe a boat ride in the harbor, or a buggy ride behind Old Dobbin ending up at Mrs. Payne's Ice Cream Parlor. I don't think you will find it dull or boring.

We have Ekstrom Brothers tandem and will ride out to Indian Lake on the new bicycle path. Built by Andrew Ekstrom of cinders, lime and clay, five feet wide and as smooth and hard as pavement, it winds through the picturesque woods to the Bee Hive built by Hiram Quick—it looks exactly like an enormous bee hive, and is used as a refreshment stand and for dances now and then; the grounds are ideal and very popular for picnics. An old Catholic Mission, the first church in this territory, stood on this point, and, according to my copies of the original Government Surveys dated 1845, there was an Indian village about where Arrowhead Inn stands, Indian gardens and an Indian clearing on the present golf course and a potato field across the road from the LeDuc residence. An Indian sugar camp occupied the present Fox Farm, and there was an Indian lime kiln just north of Maple Grove School.

How did we raise the money to build the path? Those interested got together, raised the money and hired Andrew Ekstrom to build it. Some of us put in labor instead of money, and it only took ten days.

Nowadays, public projects are usually sponsored by some club, which appoints a committee, studies it, discusses ways and means and sometimes it is possible to get state or federal aid. That seems absurd. First, some club members will not be interested in the project, and some folks who are interested will not be club members. You drag the dead weight of uninterested club members without the help of many nonmembers who are interested. The way to get things done is to get together all who want it done. That was one advantage we had before small community life was hog-tied by so much regimented confusion. Second, we hadn't reached the humiliating stage of asking state or federal aid for strictly local projects.

This is the age of the bicycle craze. Everybody has wheels in the head if not on the ground, and what to wear when wheeling is a leading topic of conversation among the ladies. It is improper for a girl to show her "limb" above the ankle and there are many different styles of bloomers and divided skirts designed to solve the problem.

The Rolls Royce of bicycles is the Columbia, costing $150. Carl Ekstrom has one. I have a fourteen-pound Thistle that cost me $110,[51] and Carl Klagstad, a very speedy rider, has a twelve-pound Thistle which he uses in century runs—road and track races around Chicago.

We will stroll over to the Star Opera House to the first Schoolcraft County Fair. The Granges have well-decorated booths and splendid exhibits of what this north country can produce. (Back in the 1870s a man was ridiculed for planting some potatoes because any fool would know potatoes won't grow here.)[52]

The whole community gets together at the Star Opera House, because community life has not yet been clubbed to death by dozens of little self-contained clubs and circles, each lady entertaining her bridge club, and the children, lacking home and family life, making one of their own, and committees of men who individually can do nothing but as a committee can get together and decide that nothing can be done. I will get tickets for "H. M .S. Pinafore," which the teachers and local talent are staging next week and I promise you it will be good. The Star Opera House was the center for all community activities: political rallies, caucuses and elections, bazaars, lectures, Lyceum courses, barn dances, religious revivals, charity balls, amateur theatricals, roller skating, baseball and professional road shows—until John Hancock and I built the Manistique Opera House, now the Oak Theater, in 1903. There is always something

Linderoth Studios; W. S. Crowe collection

Manistique Indoor Baseball "City" Team. Michigan Champions 1902-07. Top row: Harley Garland, right field; Vic Deemer, third base; John Forshar, second base. Middle row: S. Dean Crowe, left shortstop; J. C. Wood, first base; Ben Gero, Sr., manager; Ed Jewell, left field. Bottom row: F. Paulson, left short; William S. Crowe, pitcher; J. Williams, catcher; F. Guinan, right shortshop.

doing at the Star, and the way some of these old folks can throw their weight around in square dances, and what some of the local clowns and clog dancers can do with an old dish pan, iron spoon, turkey feather duster, tin horn and a bit of calico is something to see. The first graduation exercises of the Manistique High School will be held soon at the Star. Oren Quick and Pem Tucker comprise the class of two.[53]

The Star Opera House is now part of the Manistique Garage. The unobstructed floor space was fifty-three by ninety feet. The original wooden roof trusses support the ceiling in which we can still see the mouths of forty-two empty beer bottles set bottom side up with their mouths projecting downward through the ceiling to improve the acoustics for lectures, graduation exercises and other functions. Ernie Carlson told me about these bottles. I had never heard of anything like that, so I got William Stephenson, who works at the garage, to climb up with me into the space between the ceiling and roof peak, and we managed to unfasten a couple of them. It was as hot as a Dutch oven up there and a very dirty job, with the accumulated cobwebs, dust and grimy dirt of over seventy-five years. The names of two prominent Menominee businessmen, Leisen and Henes, are stamped in one of the bottles. They were put on display in the Manistique office of the *Daily Press*.

Once in an indoor ball game in the Star Opera House, a visiting player batted what would have been a sizzling home run straight down the center of the diamond, but it struck one of those overhead trusses, bounced back and was caught by our catcher, Jack Williams—the only time I ever heard of a batter being robbed of a home run by the opposing team's catcher.[54]

GOLD, SILVER AND GO-BETWEENS

I noticed when you bought those tickets you gave the clerk a peculiar yellow coin about the size of a quarter and he gave you back four big silver dollars in change. What was that? I never saw anything like it before," said my young companion.

My dear girl, I never realized that you are seventeen years old and have never seen any real money. That was a five-dollar gold piece. Gold, and silver to a limited extent, is the only real money there is. Gold has been accepted as money from time immemorial. To paraphrase Tennyson's *The Brook*: "Governments may come and governments may go, but I go on forever." Our government called in all the gold about the time you were born and stored it at Fort Knox in Kentucky, but don't believe any of those wisecracks about digging gold out of one hole in the ground to bury it in another, because all that gold at Fort Knox is really working overtime. Money is just a convenient go-between commodity, which makes it possible for you to exchange your goods or services for the goods or services of others. It must have certain qualities to make it suitable as a go-between commodity and gold possesses those qualities more than any other commodity. It must also have a stable value. The farmer who sells his potatoes today and wants to buy something next year must know that the go-between commodity he accepts for them will be worth as much next year as it is now.

You see, the farmer and the bank clerk never work for the go-between commodity, money. They work for, and trade their produce or services for, other things—groceries, clothing and other goods. They wouldn't take money at all if they couldn't exchange it for other goods. All of which means that there must be production of other goods before money can be worth anything.

If the amount of other goods produced is less than the value of money (including all promises to pay money, measured by the gold standard) outstanding, then you have inflation of money to the extent of the difference.

It is perfectly silly to think we can cure or stop inflation of money by marking the price of the goods left on the shelf up or down. In fact, arbitrary price control of goods to stop inflation of money works exactly opposite to the way it is supposed to work. It stops production of goods which might make the inflated money good.

The money in free circulation in Manistique in the 1890s was gold, silver dollars, gold and silver certificates, National Bank notes secured by U.S. Savings Bonds, United States notes—Greenbacks—and C. L. Co. coupons good for merchandise in any store in the county.

It is the custom for workers to save part of their wages. The bank pays four percent interest, the C. L. Co. issues six percent Certificates of Deposit, and usually when a bank depositor saves enough ahead he will look for a good investment in a seven percent mortgage, or a note. School Orders, payable when the schools get the money, are a good investment at eight percent, and time checks and camp orders can be bought at varying discounts. Every winter the C. L. Co. issues a lot of checks payable April 15 and May 1 to log jobbers, if necessary, and for timber purchases. Then in the spring Mr. A. J. Fox borrows from the Detroit banks against the summer shipments of lumber to take up all these checks.

ALL-DAY PICNICS

Mr. J. D. Mersereau, Secretary of the C. L. Co. and Superintendent of the Presbyterian Sunday School, has chartered a special train to the Big Spring for the annual picnic, and we will go along. Sunday School attendance has been perking up for several weeks; the children have been waiting expectantly, and it

An early raft on Kitch-iti-ki-pi (Big Spring), circa 1890s. The spring is sixty feet deep and four hundred feet across.

Henson Printing Co.,Grand Rapids; *Manistique Pioneer-Tribune* was announced last Sunday that the picnic this year would be at the Big Spring. We are due to leave the church about 8:30 A.M., and the train leaves from the depot on the west side at 9 A.M.

The whole congregation will be there, because it is an all-day affair and one of the high spots of the year. Each family packs a big picnic basket, and you should see the stacks of cold fried chicken and ham sandwiches, baked beans, meat loaf, creamed potatoes and squash, boiled eggs, cheese and sardines, salads, pickles, jam and jellies, home baked bread and rolls and sugar cookies. There are apple, peach, rhubarb and pumpkin pies and sunshine, chocolate, sponge angel food, layer and spice cakes. We will have ice cream, lemonade, hot coffee, milk, candies, nuts and fruits. No one goes hungry at the Sunday School picnic.

The men will be busy gathering wood and making fires for the women or sitting around swapping tall stories about the good old days before the railroads came.

There will be all kinds of contests, games and sports for everybody: baseball, horseshoe pitching, potato and sack races, three-legged and backward races, jumping contests. The women will take part, too, except perhaps in the three-legged race. There is a raft of railroad ties and boards on the Spring, big enough to take four or five people out at a time.

Everybody has a grand time, especially the kids—the kids under sixteen and over sixty. Those folks in between take things too seriously and worry too much about the kids. There will be lots of singing and accordion, mouth organ and guitar playing, especially on the train coming home. The Baptists are having their picnic next week in Miller's woods at Harrison Beach, and will go in six or seven big wagons with hay racks.

"I never knew there was a railroad from Manistique to the Big Spring," said my young companion.

Oh, yes. The train runs over the Manistique & Northwestern to South Manistique. then along the shore of Lake Michigan to Thompson, and from there to the Spring over the Delta Lumber Company's railroad. The only other way is by Fuller and Maclaurin's or Bebeau's steam launch up the Indian River and across the lake and then hike a half mile through the woods.[55] Railroads are the big thing nowadays.

Mr. Weston planned to build a railroad along the west bank of the Manistique River and dump his logs in at the upper mill and north of the Soo Line bridge, but the Soo Line wouldn't stand for it because so many log railroad crossings (there were already three) over their regular lines were becoming a nuisance. In those days if one railroad wanted to cross another one, it usually had to, like a famous Civil War raider, "git thar fustest with the mostest men." Frustrated in his plan, Mr. Weston bought Hall & Buell's road which ran from South Manistique to their Manistique River pull-up, and then built the Manistique & Northwestern in 1896 from Manistique to Shingleton at a cost of $250,000. It made a profit of $200,000 the first year hauling C. L. and Weston Lumber Company logs to the dump on the Manistique River at the same rate per thousand that it was costing to drive the logs down the Indian and through Indian Lake. The C. L. Co. operated this road until about 1902, when they sold it to Dan Kaufman of Marquette for $500,000. Kaufman reorganized it as the Manistique, Marquette & Northern, increased the capital and built a large car ferry which ran between Manistique and Northport for some years.

COLORFUL CHARACTERS

"Who are these rough-looking men we are meeting?" asked the teenaged miss.

Those are lumberjacks or river drivers. They are rough and tough, physically, but they won't harm you. You would be perfectly safe in a crowd of them, because if one of them bothered you, the others would take care of him.

"And those two men with packs on their backs and little axes in their belts?" she continued.

I told her they were Jo Chenord and Jim Richardson, two of the C. L. Co.'s landlookers, or timber cruisers. The company has four or five of them out most of the time. They cruise timber and can tell very closely how many feet of lumber there are in a forty-acre tract. They also blaze boundary lines so camp foremen will know where to cut.

More questions. She does ask a lot of them.

"Who is that old man on the corner in the fur coat, and what is that in his hand?" she asked.

Oh, that's Frank Jachor, the Village Marshal, and that is his ear trumpet. He is very deaf.

"How funny, a deaf policeman with an ear trumpet," she said.

Yes, but don't underestimate him, like the two lumberjacks who found themselves in the city lockup with very sore heads when they woke up. He is deaf, but he sees and knows everything that is going on. He is as strong as an ox. and that club he carries is not for ornament. He is part Indian. How does that strike you? An Indian to tame the savage white men.

I had two experiences with Jachor myself. At the time of the big fire in October of 1893, Mr. A. S. Putnam asked me to watch some goods they had carried out of the drug store and piled in the street. Jachor came along and saw me and said, "You get over on that pump." I started to explain, but he seized my arm (I can feel his grip yet) and marched me over to the fire pump, a hand engine with about twenty-four men pumping up and down. I was in dutch with everybody, including the spectators—one of whom afterwards condescended to marry me. The men on the pump called me a "no good loafer," and Mr. Putnam thought I was "undependable."

Often, when I traveled outside, folks would ask me where I came from; when I said Manistique, if they had ever heard of such a place at all, they would laugh and say, "Oh, yes, that town with the deaf policeman and the hotel with the long name nobody can pronounce, and the Irish cook who can plank whitefish like nobody's business and who once chased the hotel owner out of her kitchen with a whiskey bottle." Now I wonder how they knew all that!

At another time—it happened to be Halloween—Art Graham and I were walking down Maple Street minding our own business when we saw a gate lying in a yard inside a board fence. We had noticed that Rev. Rooney's gate was missing when we came by there. Unluckily for us, Jachor happened along and saw us looking at the gate. "You fellows take that gate right back where you got it from," he said. We started to explain, he started for us, we started for the gate and meekly carried it back two or three blocks and hung it up on Rev. Rooney's fence.

INDIAN TOWN

We will take a walk through Indian Town in the sand hills on the west side. The only streets are walks of wide boards laid on the sand, and the wooden shacks are built

M.W. Nuss, photographer; Jack Orr materials

Indian Town was located on the sand dunes east of present-day Manistique. In the 1880s-1890s, many immigrants from Sweden, Finland and England lived in Indian Town, along with Native American families.

haphazard and facing in every direction. There are many Indian children playing about, and the women are busy making baskets, moccasins and bead work of all kinds. The C. L. Co. has about seventy-five Indians on the payroll, and they are good workmen for the most part. We call the Indians "savages," but the older I get, the less certain I am of that. In fact, I sometimes question just who were the savages. The Indians who met Columbus were friendly and thought the white men were gods and supposed the Spaniards to have come from heaven. Barnes' *History of the United States* says, "How sadly and how soon these simple people were undeceived."

The Indians in the Manistique area had a village on Indian Lake, a sugar bush, a lime kiln, a potato patch and undoubtedly they grew corn and some other things. They hunted and fished for food only, never for sport. I am not advocating that we adopt their way of life, but I do think we should understand that primitive people are not necessarily "savages."

LOCAL POLITICS

We will go back to the Star Opera House, where everything of general community interest takes place, and look in on the Republican Caucus for nomination of officers. The atmosphere is so thick with cigar smoke you could cut it, and big brass cuspidors stand around here and there. The old-timer who can't hit one of these at fifteen feet just doesn't belong.

There are eight hundred to a thousand men on the floor. The only furniture is a long table down in front and a few chairs for the tellers, because nobody ever sits down at the caucuses.

There is nothing secret about the voting. Slips are printed in advance; each leader passes them among his followers. The tellers pass through the crowd with tall silk hats into which the voters drop their ballots. When all have voted, the contents are dumped on the long table and the tellers, with the leaders scrutinizing closely, sort and count them. The chairman announces the result, the slips are dumped in the waste basket, and that is that. Everything open and above board—maybe.

No women. Although women had begun to show signs of intelligence, they hadn't yet reached the point where they could be trusted with the ballot. The creatures couldn't even smoke yet.

Nomination is tantamount to election because every man prominent in the community except perhaps five or six Democrats led by Charley Coon (always so busy that "busy as Charley Coon" became a village byword and is still heard now and then) will be there with their followers: Denny Heffron, suave and smooth and a very likable fellow; George Holbein, editor of the *Pioneer-Tribune* (the opposition paper always referred to him as "Bean Hole"); W. C. Bronson with the Methodists and church people behind him; T. J. MacMurray, editor of the *News* (and grandfather of the screen star, Fred MacMurray); Charley Mersereau, with a personality that would charm a wooden Indian; George Frankovich and George and Frank Lasich, with their Austrians, Poles and Hungarians; Nels Johnson, with his Scandinavians, who said: "Yah, he bane gude faller. I vote like he say." Not quite all the Scandinavians voted "like he say," however. Also Ben Gero and W. F. Crane, rival lumber inspectors and politicians; Walt and Fred Orr and Frank Cookson with their Yankee drawls and lumberjack followers; Patsy Mack—an Irishman out of politics is like a fish out of water. My dad's parents came from Ireland and I used to get a great kick out of listening to Patsy Mack and Rev. Rooney, the Baptist minister, who were great friends. The above were only a few of the many colorful characters attending these caucuses.

These overgrown village caucuses continued until Manistique became a city in 1901 and voting precincts were established in four wards, each ward electing two aldermen and all wards together electing a mayor; this system continued until the city adopted a charter and the city manager plan in 1925.

Manistique's Fourth of July celebration at the Fairgrounds, 1914.

From a postcard; Superior View Studio, Marquette

Schoolcraft County Historical Society

The circus comes to town. Elephants and a water buffalo parade through downtown Manistique in 1910.

Organizations of political influence were the GAR, the various lodges, the churches and the saloons. The "company" (C. L. Lumbering Company) exercised a large influence but a company ticket would have been defeated, even though ninety-five percent of the community's business depended on the company.

This was America all over. Politics were strictly local, and Washington or Lansing had nothing to do with local politics or the affairs of individuals, their job being to provide national defense and security. Nothing like it existed in Europe, where the only parties were the rulers and the ruled. And, when folks talk about the terrible danger and the war of survival we are in, and especially about the way our war secrets are being stolen, I think of these words, Kipling's, I think:

They copied all they could follow,
but they couldn't copy my mind,
And I left 'em sweating and stealing, a
year and a half behind.[56]

I have a lot of faith in the American boy's resourcefulness and initiative, mechanical know-how, and background of 150 years of freedom; if we can only keep in the American Way on which Washington, Jefferson and Franklin started us, I do not think we have much to fear from any outside source.

ECHOES OF GETTYSBURG

The Fourth of July was heralded far and wide for weeks in advance by big colored posters, decorations and other means, and was ushered in by cannon salutes at sunrise. Nobody slept from then on. Small boys, and some not so small, were deliriously happy with every conceivable noisemaker, and what they couldn't think of in noise-making gadgets hadn't been invented yet. Torpedoes and fire crackers—no limit to size—were going all day long. One Fourth I invested in two twelve-inchers and blew three pickets off Mrs. Fuller's fence with one and exploded the other under an old tin pail, which sailed away in the air and rolled off the housetop and down the street—my modest contribution to the general racket.

The cannon salutes, the GAR rifle salutes over the graves of Union soldiers, the torpedoes and firecrackers were like the sound of musketry and one could almost imagine a battle in progress. In fact, sham battles were sometimes staged on the hills near Lakeview Cemetery. The old army 45-70-500 Springfields with black powder and blank cartridges made plenty of noise and smoke, which, punctuated with camp fires here and there, gave faint imitations of Gettysburg, Antietam and Chickamauga on a small scale, and our imaginations did the rest.

The whole community and countryside gathered early for the giant parade headed by several brass bands, the GAR and the W.R.C. (Civil War memories were still fresh in those days). They paraded to the cemetery where appropriate exercises were held and salutes fired, and from there to the Central School yard or Court House square, where the orator of the day read the Declaration of Independence (to which everyone listened attentively), and twisted the British Lion's tail with everything he had, to the great satisfaction of the crowd. Refreshment booths, where the various Ladies Aid Societies served good meals for twenty-five cents to fifty cents, and ice cream stands bloomed all over the place. In the afternoon there were baseball games, field sports and bicycle or horse races at the Fair Grounds, topped off at night by as large and spectacular a display of fireworks as the community could afford. There were also many private displays all over town, and it was well after midnight before it was again "all quiet on the Potomac" (or the Manistique).

CHAPTER SIX

The Froth and the Truth

fantastic and unrealistic portrayal of life in the Michigan pine lumber days in some recent literature gives this generation a greatly distorted picture of those times. The old lumberjacks are pictured as a red-eyed, hell roarin' rough and tumble, fighting gang, jumping on opponents' faces with calked boots, thinking only of whiskey and women, taking the town apart when they came down out of the woods. We are told that decent women kept off the streets when the lumberjacks hit town, and that the camp boss had to be a tough, high-rollin' Bull of the Woods able to lick any bully in camp to maintain his authority.

It wasn't like that.

Part of my job as bookkeeper was to keep the payrolls and settle with the lumberjacks when the camps broke up. This was during the seven-year period, 1893-1900, when the industry was operating at peak load in northern Michigan and the Chicago Lumbering Company and the Weston Lumber Company were cutting approximately a hundred million feet of white and Norway pine every year. They employed about fifteen hundred men, including the independent loggers, jobbers and farmers who put in some logs for the Chicago Company every year.

"The old-time woodsmen hugely enjoyed filling up the greenhorn with fancy stories. . . ."

I knew all the camp foremen very well, and most of the lumberjacks by face and name, and was well acquainted with many of them. They were mostly young men who worked in the woods in winter and in the mills and yards or on the river drives in the summer; they saved more or less of their wages and eventually got married and settled down. Among them were John Falk, August Carlson, John Blomquist, Harry Adams, Charles Peterson, Axel Victorson, E. A. Palmquist, George Frankovich, Jake and Tom Edwards, the Woods, the Grahams, the Ekstroms and many other well-known citizens.

Just like a bank, the C. L. Co. issued six percent Certificates of Deposit in which many of the employees invested their savings. I remember renewing a $3,500 CD for George Frankovich year after year and paying him $250 interest. Two hundred and fifty dollars of those Old Deal dollars would go far toward supporting a family in those days. George and Frank Lasich also had large deposits, and Ab Orr, the company barn boss, had over $10,000 in company CDs.

Among a thousand lumberjacks there were perhaps twelve or fifteen pretty tough guys with whom it was just as well to be careful, but even these usually confined their fights and brawls to the saloons. I never had the slightest trouble settling up with any lumberjack; I never saw a street fight of any importance, and the story that women kept off the streets when the lumberjacks hit town is all poppycock. The attitude of the lumberjacks, even when drunk, toward decent women was chivalrous and respectful, and the women came and went as usual when the camps broke up. Of course, no decent woman ever was seen in a saloon in those days.

Perhaps forty or fifty of the boys would blow in or be rolled for their stakes, have what they called a good time for a couple of weeks, then go up on the drive for the summer, have another spree in the fall and go back to the woods for the winter.

The twenty-seven saloons in Manistique did a rushing business in the spring and fall—I don't mean to say that most of the lumberjacks never took a drink. Most of them did now and then, but on the whole the overall picture of the lumberjack is about the same as for other men. The lurid stories published by sensational writers are based on the doings of not over two percent of them and the carryings on of the few are held up as a true picture of the pine lumber days, when, in reality, they were only the froth.

It is easy to write a sensational story about the froth and turmoil that activates the surface of all things, but all one can do with the great mass of underlying truth is to present the facts in as readable shape as possible. Sometimes truth is stranger than fiction.

I knew P. K. Small very well. He got his deformed nose in a fight, I was told, but I never saw or heard of him biting the heads off snakes, birds and toads for a drink, or that some other lumberjack bit P. K.'s nose off because P. K. had bitten the head off the other fellow's pet owl, until I read these tales in a book written by an author who wasn't born until three years after the last of the big pine had been cut and the C. L. and W. L. companies had quit operating. Frank Cookson told the best story about Small. P. K., who had a fine education, was convicted in Trout Lake for being drunk and disorderly, and when Judge Steere asked him if he had anything to say, P. K. said, "Judge, I've been arrested for every crime in the book, I guess, except murder, but being drunk and disorderly in Trout Lake caps the climax. I have nothing to say."

From the pictures painted of Seney, one might suppose the town was nothing but a collection of saloons and dives populated by a rip-roaring fighting gang of hell-raising lumberjacks and prostitutes. On the contrary, there was a church, several stores and a school; the majority of Seney's inhabitants were decent, respectable people; and Trout Lake was considered by the lumberjacks themselves to be a much tougher town than Seney.

LOG PIRACY

My younger daughter had a hobby for collecting odd chunks of driftwood and came to me with a peculiar piece she found on the shore of Lake Michigan and asked if I could tell her what it was.

"Sure," I said. "It's part of a big white pine log that was stolen from the Chicago Lumbering Company about fifty-three years ago. If you'll find the rest of it, I'll buy you a nice new dress."

"Don't try to be funny," she said.

"I'm not being funny. It's very simple. That faded 'hieroglyphic' is the C. L. Co.'s log mark, and anyone can see that this is part of a slice about one and a half inches thick which was cut off the end of a log. The only reason anyone would have for sawing the end off a big log in those days would be to get rid of the log mark and substitute his own. It happened over fifty years ago because there were no other operators in big pine in C. L. territory after 1900. After 1900, the C. L. Co. was bringing in most of its logs by rail from above Steuben. And, from its appearance, this piece of log has been knocking around in the water for at least fifty years, so the tree probably grew somewhere along the Manistique River. This is part of the outside of the log, and the center is missing, but by extending these cracks which converge towards the center, we find that they would intersect just nineteen inches from the circumference; therefore, the log was thirty-eight inches in diameter at the small end, as logs are always scaled at the small end. Trees have what are called 'medullary rays' which radiate from the center to the outside, and cracks or checks in the ends of logs usually follow these rays. This tree would be around 140 feet tall; by counting the rings and estimating the number of rings in the missing center—bearing in mind that the center rings would be at least four times as wide as those next to the bark—I think you will find that the tree was over three hundred years old."

(An actual count afterward verified that the tree was about 280 years old.)

This could have happened another way. A crooked jobber putting in company timber on contract could have cut the end off after the company's scaler had measured it, so on his next trip the scaler would scale and mark it again, thus giving the jobber credit twice for the same log.

There was some timber and log stealing in the 1880s and 1890s, but it was rare, and it was among the smaller and more irresponsible owners and jobbers.

The impression has been fostered by sensational writers that the old-time lumbermen were "robber barons" and "exploiters" who grabbed timber land and logs whenever and wherever they could from the government, state or any other owner; that their only thought was to cut out and get out; and "to hell with the country" and "the public be damned." But it is simply not true.

They would imply that the Stephensons, Goodmans, Ludington and Upham of Menominee and Marinette; the Hackleys of Muskegon; the Wentworths and the Fordneys of Saginaw; the Detroit Whitneys; the Merrills of Wisconsin; the Walkers, Smiths, Shevlins and Weyerhausers of Minnesota; John M. Longyear and William G. Mather at Marquette; the Westons, Wheelers, Quicks, Orrs and Mersereaus at Manistique; General Russell A. Alger at Grand Marais; and Robert Dollar (of steamship fame) at Dollarville were all little better than common thieves, which is ridiculous and absurd on the face of it.

The really big men in the industry never stole timber or trespassed intentionally on lands of other owners, because in the first place they were not that kind of men and in the second place, even if they had been, there would be no sense in stealing timber when the government was offering fine timber land to anyone who would buy at $1.25 per acre.

Nothing delighted an old cowhand on the western plains more than to stuff a tenderfoot with tall tales about six-gun men, stage robbers and cattle thieves, and the number of desperados he had helped string up. When I lived for two years on a cattle ranch in eastern Colorado in the early 1890s, if I had believed half the stories they told me I would have seen the ghost of an outlaw hanging from every limb on every cottonwood on the banks of the South Platte; but in all that time I only saw one cowboy with a revolver and he was practicing on jackrabbits.

Likewise, the old-time woodsmen hugely enjoyed filling up a greenhorn with fancy stories, especially if he was a writer gathering material for a book, in which case they went all out and outdid themselves inventing Bunyanesque tales about the "red-eyed, bearded, fighting, swearing, savage, high-rolling, whoring, whiskey-drinking, quarrelsome and tough lumberjacks and whitewater men" of the Gay Nineties.[57]

Actually, the good people in any community so far outnumbered the bad that the old fellows, for mere lack of material, just had to draw on their imaginations—very good ones, by the way—to make up the exciting stories the younger generation seems to require.

The tragedy of all this, however, is that it leaves the present generation with a completely false and distorted picture of real life in pioneer days.

A REBUTTAL

I now quote, and will try to refute, some passages from a book, *Call It North Country*, written by a gifted young author who wasn't born until three years after the C. L. Co. had ceased operating but whose sponsors assure us ". . . has fished, hunted and traveled all over that land he writes about, has studied its history and talked to its people."[58]

I QUOTE: "The jacks, starved for liquor and women, overran the sawmill towns like pillaging conquerors. . . . Manistique, Escanaba, Menominee—the cities were abandoned to them. The 'good' people stayed indoors for days, from the first onslaught until the last drunk was plucked from the gutter. . . . they jammed the broad straight streets from sidewalk to sidewalk, a surging flood of bearded men, fighting, drinking, whoring. . . . Up and down Ludington Street in Escanaba they swaggered . . . Manistique became the seat of Schoolcraft County and nerve-center of the lustiest, bawdiest logging operations in Upper Michigan."

And speaking of Seney, "Hell Town in the Pine," the author says, "This God-forsaken hamlet in the pines housed the offices of a half-dozen of the largest lumbering companies then operating in Upper Michigan, including the famed C. L. Company and Alger, Smith." The book goes into sordid details about conditions in Seney which I would not care to repeat in print. The truth is, that since the time of Adam and Eve and David and Bathsheba until the present day, wherever there have been men there have been women, and some sin. It has always been so. It is so now, and it will always be so, probably until the sound of the last trumpet. That is life as it is. Everybody knows it, and there is no necessity for writing a book, or going into detail about it.

I REFUTE: The statement that the cities were abandoned to the lumberjacks and the "good people" stayed indoors and that women kept off the streets when the camps broke up in the spring is pure baloney. Anyone who doubts this can easily confirm my statement by simply asking any woman who lived in Manistique in the 1890s. I am sure they will agree that the women came and went as usual when the camps broke up and that the lumberjacks did not molest decent women.[59] Furthermore, the camps did not all break up at the same time, and lumberjacks were drifting into town over a period of three to five weeks in the spring. I was keeping the C. L. Company's books at the peak of their operations in the 1890s and I never saw any such scenes as are described in this book, nor did I ever have the slightest trouble settling with the lumberjacks when they came down in the spring.

We had a branch office at Seney in the charge of Mr. A. C. Carpenter, a prominent member of the Methodist Church in Manistique. He brought his books and records to me monthly for check-up; if Seney had been half as bad as depicted, he certainly would have said something to me about it. Seney was a rough frontier town, no better or worse than dozens of the frontier towns in the lumber and mining districts of Michigan.

I deny absolutely the implication that Manistique was the "lustiest, bawdiest" town in Upper Michigan and I assert that on the contrary, it had the best moral atmosphere and character of them all. Anyone who lived here through the early days and was acquainted with the personal character and integrity of its owners and founders knows this is so. About the first thing the Quicks, Orrs and Mersereaus did when they came to Manistique was to start Sunday schools and help organize churches.

The book *Call It North Country* had this to say: "George Orr was a lumber king who determined that evil should not flourish in his realm. To that end, he and his company bought up most of the land in and near Manistique. Saloons and bagnios were denied housing; Manistique would be an island of purity amidst the hell towns in the pines. But Orr's agents overlooked a small parcel of land in the west end. Canny Dan Heffron snapped it up and opened a saloon, and over his saloon he maintained what have been called 'club rooms.'. . . Other free spirits joined him; soon the land that Dan Heffron had grabbed became the chief business section of Manistique. George Orr and his do-gooders were undone." And "George Orr, thwarted at Manistique, would have loved Hermansville. The good people got the upper hand in the beginning and they never let go."

THE FACTS: In 1893, when I came to Manistique, the saloons were confined mostly to the "flatiron" point between Pearl and Water Streets entirely outside the "chief business section." Dan Heffron had been arrested and convicted on a morals charge but escaped and fled for parts unknown and never came back. The Chicago Lumbering Company dominated Manistique completely for the forty years of their operations, and to say "George Orr and his do-gooders were undone" and that he was "thwarted" sounds silly to anyone who lived here in those days.

Furthermore, it was not George Orr, but the Wheelers and J. D. Mersereau who were principally responsible for this clause in the C. L. Co. deeds: "It is expressly declared . . . that said premises shall never be used for . . . manufacturing, storing or selling intoxicating liquors, whether distilled or fermented, nor for a house or place of prostitution or assignation, nor for any business or occupation prohibited or punished by the law of the land; and if said premises or any part thereof shall at any time hereafter be so used, all rights . . . shall thereupon cease and be forfeited absolutely and forever, and said land and appurtenances shall thereupon revert . . . to said party of the first part."

As to whiskey in the camps or on the drives, George Orr was an absolute czar. The lumberjacks knew that any man caught with a jug or bottle on the job would never have another chance. In fact, most of them would have reported an offender, because no one knew better than the lumberjacks that, although whiskey and women might mix off the job, whiskey didn't mix with double-bitted axes, crosscut saws, falling trees, widow-makers, icy roads, log jammers and river driving. Lumberjacks worked in gangs of three, and one man drinking could jeopardize the whole gang.

Mr. Quick was just as adamant with the mill and yard men. In fact, no employee in any capacity who mixed his drinks with his work could work for the C. L. or W. L. companies, which was a factor in their remarkable non-accident record.

The only "bagnio" operated near Manistique was a boarding house known as the Klondike, on the old State Road to Garden near its junction with present M-94. The line of demarcation was sharply defined in those days. The ladies from the Klondike dressed in the height of fashion, usually traveled in pairs when shopping in Manistique, never spoke or were spoken to by anyone on the streets and were rarely seen entering a saloon. It was they, rather than the "good people," who really needed protection.

Shortly after the First National Bank opened in March 1900, a well-dressed lady came in and made a deposit. Mr. Teeple, assistant cashier, waited on her, and then asked me, "Did you notice? That was Madame ____________, from the Klondike. She rolled her stocking down and pulled out a roll of bills that would choke a horse." We will not go into further details on this subject, as seems to be the modern fashion, but, like the Gibson Girl of the Gay Nineties, will leave something to the imagination.

TO QUOTE FURTHER: "In the 70's the real boom started. . . . 'Bijah Weston came from New York with his Weston Lumber Company, and George Orr formed his powerful Chicago Lumber Company; they battled sin together and made fortunes. The Jamestown men came in from New York, and Robert Dollar logged without notable success, gave his name to the town of Dollarville, and went on to the coast and a fortune in shipping. . . . great lumber companies sprang up almost overnight . . . they cut out, and got out, leaving desolation behind them, and they cared not a damn for legalistic hair-splitting (to 'log a round forty' meant to cut the timber all around a forty-acre tract, and might involve cutting a lot of timber that belonged to somebody else). Among the great companies were the I. Stephenson Company, the Bay de Noc Company . . . the Alger, Smith Company and the Wisconsin Land and Lumber Company. . . ."

All of which amounts to a direct accusation that the really big men in the industry were lawless thieves, utterly regardless of the rights of others. A more complete collection of misstatements could hardly be assembled.

'Bijah Weston did not come from New York with his Weston Lumber Company. George Orr had nothing whatever to do with "forming the powerful Chicago Lumber Company." He was a private soldier in the Union Army in 1863 when the company was organized. The "Jamestown men" did not come from New York. Robert Dollar made a fortune logging in Michigan[60] and on the Pacific coast before he went into the shipping business at the age of fifty-seven. (Mr. William Wheeler once invited me to lunch at the Merchants Club in San Francisco, where I met Capt. Dollar. A finer or more rugged man never lived. The Chinese preferred his word to most men's bond.) I never heard the expression "to log a round forty" until I read this book.

"Forgive us our trespasses as we forgive those who trespass against us" calls for a mighty lot of forgiving on the part of the big lumbermen whose lands were trespassed upon a dozen times for every time they trespassed unintentionally upon the lands of others. Absolute proof of one instance is the piece of a big pine log stolen from the C. L. Co. now on display in the Manistique office of the *Daily Press.* The slanders of writers who have labeled them as robber barons and timber thieves also needs some forgiving.

I have no quarrel with this young writer. He is entertaining, typical of a whole school of professional writers who can write but don't know what they are writing about, and so have to depend on hearsay, and who have overloaded the shelves in the past twenty years with a whole flood of trash on technical subjects, mostly economics. They should stick to pure fiction instead of camouflaging it under an assumed title.

When passing judgment on the lumbermen who cut down the old pine trees and left us a heritage of stumps and sand plains, we must consider the circumstances under which they lived and not judge them according to conditions obtaining sixty years after they passed away.

The era from 1840 to 1900 was, barring the interlude of the Civil War, an era of great expansion in industry and population in the United States. Millions of immigrants, with little, if any, restriction on their numbers or nationality, were pouring into the country from Europe to find new homes and opportunity in the land of liberty they had heard about, and they were absorbed almost immediately into the American Way. And that fact brings up a thought. If fifty million free Americans weren't afraid to admit just a few thousand foreigners from 1880 to 1890, why are 150 million Americans so afraid to admit just a few thousand aliens today? There is plenty of food for thought here.

Now, millions of those foreigners and millions of Americans were swarming to the great Middle West, following advice of Horace Greeley, "Go West, young man," to make homes and build farm houses, barns, fences, silos, cattle pens and corn cribs, and grain elevators, railroads and telegraph lines on those great fertile prairie states destined to become the Bread Basket of the world, all of which required billions of feet of lumber, ties and posts—and there were no trees on the prairies.

At the same time the whole state of Michigan was an unbroken forest of white and Norway pine—the finest building material in the world—interspersed with thousands of acres of cedar, tamarack, spruce and fir. In short, billions of ties, posts and poles were all accessible by easy water transportation on the Great Lakes, right to the prairie farmer's doorstop.[61]

So the old-time lumberman, thinking just as much, perhaps more, of future generations than we do today, cut those trees and shipped the lumber and railroad ties to the prairie state settlers, and instead of despoilers were actually builders, helping to develop our country's resources and making homes for millions of people.

Now it seems that in the minds of some, Messrs. Weston, Fox, Orr and Quick and their contemporaries in Muskegon, Saginaw, Ludington, Alpena, Marinette and Menominee shouldn't have cut those trees. They should have left them for us. (Which generation of "us," I wonder.) They should have let the trees stand, and let the farmers of the great prairies travel and live in their covered wagons, or tents, or maybe burrow in the ground like the prairie dogs, gophers and coyotes.

On a shelf in the C. L. office I had some shoeboxes in which I filed the various camp reports, log scales and van orders, which the tote teamsters brought in to be checked by Mr. Orr (then Vice President and Woods Superintendent of the C. L. and W. L. companies) before turning the orders over to the stores to be filled. It was Mr. Orr's custom to come over to the office after supper with an old army overcoat slung over his shoulder (I never saw him wear an overcoat any other way) and inspect the scales and reports. Quite often he would scratch items from the van orders.

Being a youngster, I rarely spoke to Mr. Orr, Mr. Quick or Mr. Mersereau until they spoke to me. However, one evening Mr. Orr seemed to be in a cheerful mood and I ventured to say to him, "Mr. Orr, don't you think the company would make more money by saving more of the timber for the future instead of cutting so heavily now?" And he said, "Yes, Will, that is probably true, but there are a lot of things we would like to do if we could only stop eating for ten years." I had no answer for that.

I have two wonderful daughters who are in complete agreement and harmonious accord on one subject—Dad. "Dad shouldn't do that. Dad ought to do this. Dad should sleep an hour after meals. Dad is just like a mule. Nobody can tell Dad anything." (To the last I take exception, because somebody told me everything I know.) Sometimes I think of that old story of the little boy whose mother said, "Mary, go upstairs and see what Willie is doing and tell him he mustn't."

The reason for this little digression is that one of these charming daughters wrote that she liked my lumberjack stories, but was afraid I might "antagonize people" and "date myself" by "running down the present." In case others might have the same idea, I will say that I have no intention of "running down the present." If the present jumps the track of sound and proven economics, or chases will-o-the-wisps, it will run itself down.

This is an attempt to portray a true picture of life in the horse-and-buggy days in Manistique, a town on Lake Michigan in the Upper Peninsula (and when you get the true picture of Manistique in the big pine days, you will have a true picture of every other lumber town in Michigan and Wisconsin in that colorful era). And it is based entirely on my own experience and observation, without distortion. But, to convey to the present generation a true picture and perspective of the past, comparison with the present is sometimes necessary.

And I cannot date myself by telling the truth, because the truth is never outdated. Truth is eternal and time is not a factor.

And when I say we had no unemployment insurance because there was work for everybody who wanted to work; no social security because we felt secure; no old-age pensions because families took care of the few old folks unable to care for themselves; that we had to think for ourselves because we had no commentators to think for us; that we had no governmental bureaus to plan our lives but had to plan them ourselves; that we didn't feel called upon to police the world; that we didn't fear communism or any other "ism," of which there have always been plenty; that we never received state or federal aid to teach children how to slide down a toboggan, or to build an ice rink on the sloping courthouse grounds from which the water ran off before it could freeze—I am not "running down the present" but simply stating facts.

I believe in social gains, but I also believe, and think I can prove, that we could have these and many more social gains at far less cost, and a balanced United States budget, and no fear of inflation, but that is another story.

CHAPTER SEVEN

Buying the Chicago Lumbering Company

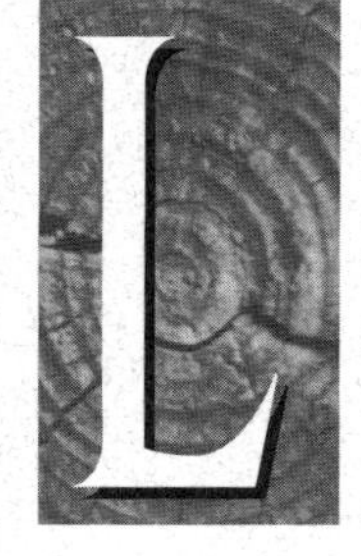

Like all things, the story of the old Michigan big pine era with its colorful lumberjacks, white water men on the river drives, picturesque sailing schooners and busy lake ports, screaming sawmills and smoky blue haze over all, comes to an end.[62]

A pattern followed by many of the old pine lumber barons, when nearing the end of their operations, was to sell out and let someone else mop up. They themselves would go west to continue operations on a larger scale, and, as they soon learned, in a vastly different environment and under more difficult logging conditions. The big pine fellows wanted nothing to do with the despised hemlock, hardwoods and short stuff; and the old-time lumberjacks looked down with lordly contempt on "cedar savages."

"You boys have a wonderful bargain and I hope you make good, but don't forget to come across with a million dollars in cash on December 1st. . . ."

The Chicago Lumbering Company crowd was no exception. They acquired extensive sugar pine holdings on the Rogue River near Crater Lake in Oregon. The Wheelers had redwood operations in California, spruce and fir in Washington and owned the evening *Telegram*, a prominent Portland, Oregon, daily. The McCormick Lumber Co., owned by Wheeler interests, operated in Washington.

N. P. Wheeler was one of the largest timber holders on the Pacific Coast, exceeded only by the Northern Pacific RR, the Weyerhausers, perhaps Crown-Zellerbach, Long-Bell and one other. The book *Pine Knots & Bark Peelers*, by W. Reginald Wheeler, mentions twenty-four lumber and timber companies in which N. P. Wheeler had interests.[63]

Weston interests also had extensive holdings on the coast, including a valuable tract of redwood and a mill near Eureka, California. In 1922, on an auto trip to the west coast, my family and I passed through Eureka, and were invited out to see a redwood tree fall. They had smoothed the ground and made a bed of brush for a cushion because the tree's great weight might shatter it. Its fall shook the ground three hundred feet away where we were standing. This tree was eighteen feet in diameter, 275 feet tall, and was five hundred years old when Columbus landed. Another still living up Yosemite way is over four thousand years old. To see it brought low in a few hours' time was rather saddening. Fortunately, a "Save the Redwoods" campaign by interested citizens succeeded in preserving large tracts of these magnificent trees, and the Redwood Highway is today one of California's great attractions.

H. C. Culver, of Corinne, Michigan,[64] became a millionaire operating at Sand Point, Idaho, largely in poles for big utility companies.

George F. Ross, of the Peninsula Cedar Co., a large operator along the Soo Line in the U.P., acquired a tract of white pine on the Priest River in Idaho in 1901, and commissioned me to go out and set up an office system for his Laclede Lumber Co. Stearns, of Ludington, operated in yellow pine at Bend, Oregon, and the Stangs of Merrill, Wisconsin, were at LaGrande, Oregon. In fact, most of the big operators on the west coast, including the Weyerhausers, the biggest of them all, came from Michigan, Wisconsin and Minnesota. Harry Thomas, a son of W. B. Thomas, secretary during the '90s of the White Marble Lime Company in Manistique, has a very responsible position in the land department of the Weyerhauser general office at Tacoma. In 1947 I spent four hours in the Tacoma office and Thomas gave me some very fine pictures and booklets descriptive of their logging methods.

The Weyerhausers at that time owned about one and a half million acres of timber land in Washington and Oregon on which there was an estimated stand of at least forty billion feet. Under their progressive cutting and replanting methods they plan to go on cutting forever; the largest sawmill cuts one million feet a day.

THE BIG SALE, 1912

How did they acquire their holdings? By purchase, some from the government, some from the Northern Pacific Railroad, from private owners, homesteaders and timber claims. In 1905 my brother, S. D. (Dean) Crowe, and I each proved up 160-acre timber claims on the Little Wind River in Washington, which we afterwards sold.[65] Going out to prove up my claim, our train, the North Coast Limited, was held up by two bandits in Yakima canyon about 7 P.M., December 16, 1905. They dynamited the safe, wrecked the express car and got away—with $40. But they were captured by a posse in the mountains before they had a chance to spend it. Crime does not pay.

In December 1911, I went to Portland to assist Dean, then manager of the Home Independent Telephone Co. of LaGrande, Oregon, in setting up a financial plan to purchase the Bell System's ten exchanges and connecting toll lines in Union and Wallowa counties. A five percent bond issue was put out, with the Harris Trust and Savings Bank, Chicago, as trustee. In San Francisco, the home office of the Pacific Co. agreed to take half its money in bonds; banks in Union and Wallowa counties took some, and the rest were held in the treasury for future capital needs.

On my return, Lou Yalomstein, then manager of the C. L. Store, came up with a proposal that we form an organization and buy out the C. L and W. L. companies.

I said, "You're joking. You know Stack (J. K., of Escanaba) had an option on it last year for a million dollars and let it drop?"

"Yes, I know, and it was dirt cheap at a million dollars, but they've had another winter's cut and I think we can get it for less, and I'm not joking. I think we can put it across."

We talked it over some more, and then went down to see Mr. Orr, who after a couple of conferences gave us the following option:

Manistique, Michigan, Feb. 21, 1912

Wm. S. Crowe
L. Yalomstein
Manistique, Mich.

Gentlemen:

In consideration of the sum of Five Thousand ($5,000) Dollars in hand paid we hereby agree to sell to you the following property on the terms and conditions as stated herein. This option to be void and of no effect if sale is not completed on or before December 1st, 1912.

The entire Real Estate and Improvements including Riparian Rights, Water Power and Dams, and Timber Contract holdings of the Chicago Lumbering Company of Michigan, the Weston Lumber Company and J. D. Weston & Co., excepting and reserving Indian Lake and River Farms, and subject to all leases and contracts now in effect.

Also to sell to you the following personal property, viz: all woods and camp equipment including horses belonging at camps, all mill and yard equipment including blacksmith and machine shops, office furniture and fixtures, abstract books, and the Manistique River Improvement Company, also all deadhead and submerged logs and other submerged forest products.

We reserve the sole use and control of mill and lumber yard to end of present season to complete this season's cut, and to carry lumber stock on yard until not later than July 1st, 1913.

We shall be entitled to all logs and cedar, hemlock ties and pulpwood cut previous to May 1st, 1912, in the water and in yards of White Marble Lime Co. at the time of execution of deeds or contract of sale. In the event that we do not saw all such logs and cedar, hemlock ties and pulpwood so in the water and in yards of White Marble Lime Co. before the shutting down of the mills for this season, you are to purchase the remainder, upon the basis of a joint estimate, at market prices.

We are to continue all our woods operations in the usual manner and our contracts with jobbers, and in the event that you exercise your option of purchase herein conferred you shall assume all contracts with jobbers and shall reimburse us for all woods expense of every kind incurred on all timber cut after May 1st, 1912.

Our stocks of General Merchandise, Hardware, Weston Mfg. Co. and Warehouse, or any of them, will be sold to you if you so desire on a joint inventory Dec. 31st, 1912, or if not we are to retain the use of buildings at a reasonable rental until July 1st, 1913, if necessary, for the purpose of disposing of the stocks.

If you complete this option we will accept $750,000.00 in full payment of the sale, the $5,000 option fee to apply on same.

All taxes for 1912 will be paid by us, and on completion of payment as herein agreed we will convey the large part of the lands by warranty deed, others which we hold on tax title or incomplete title, by quit claim deed.

CHICAGO LUMBERING COMPANY OF MICHIGAN

By GEO. H. ORR
President

By C.E. KELSO
Secretary

WESTON LUMBER COMPANY

By GEO. H. ORR
President

By C.E. KELSO
Secretary

ACCEPTED

WM. S. CROWE

L. YALOMSTEIN

We formed the Consolidated Lumber Company and completed the above option, and I will conclude this lumberjack story with a brief account of its history.

When Mr. Orr handed Lou Yalomstein and me the option to purchase the C. L. and W. L. companies' properties, he said, "You boys have a wonderful bargain and I hope you make good, but don't forget to come across with over a million dollars in cash on December 1st, or your $5,000 and whatever more you spend is gone, because that option won't be extended a single minute."

And Mr. Quick on his return from California that spring told me substantially the same thing, and added that if he were twenty years younger we would never have had the opportunity.

The price of $750,000 was for the real estate and logging equipment only. The merchandise, hardware, retail lumber, planing mill, coal and warehouse stocks would require at least $200,000 more, and logging operations would take not less than $250,000, because in addition to their own camps, they had contracts with fourteen log jobbers for over eleven million feet. Jobbers who put in over one million feet in the 1912-13 season, and whose contracts we had to assume, were:

Arthur Burton	3,519,078 feet
Frank Cookson	2,799,618
Miller & Fox	2,184,899
Julius Phillion	1,337,131

Nevertheless, although we even had to borrow the $5,000 option fee temporarily, we assured Mr. Orr that he had made a deal, and that we would have the cash on the line, which we did.

Our first step was to raise $10,000 expense money; get reliable data, cruises and appraisals, and set up a plan of organization acceptable to outside capitalists. We contacted three local men, Virgil I. Hixson, A. S. Putnam and W. B. Thomas, who, after we told them the price, put up the $10,000 expense money in return for a substantial stock bonus if, when and as our deal went through—otherwise no return. They also agreed to invest $115,000 cash in securities of our new company.

The C. L Co. was an active going concern employing perhaps six or seven hundred men at this time; neither the company nor ourselves wanted it known that an option had been given until we knew the deal was going through. Our problem therefore was to get accurate data about the business without arousing public curiosity. Mr. Kelso, Secretary, gave us a list of the lands (161,000 acres) and the city property, consisting of five general store buildings; 147 two-story houses in good repair; Ossa hotel; 650 acres of land, including all available factory sites, building lots and unplatted residence locations; docks; mills; trains; and water power. On the map, the company owned about seventy percent of the city's area.

Included in the logging equipment were ninety heavy draft horses but the only data on the timber was Mr. Orr's statement that there was "over one hundred million feet which would run over four thousand feet to the acre (he didn't consider that a stand of less than four thousand feet to the acre was a logging proposition) and where to find it. We knew Mr. Orr was very conservative and that we could depend absolutely on his word, but we would have to have more definite data to interest a Chicago or Detroit bond house. So we engaged a firm of reputable outside cruisers, who found 128 million feet of saw timber on approximately 25,500 acres and enough other products to aggregate $728,500 in stumpage values at the prices then prevailing, which was enough for our purpose, so we didn't bother to cruise the bulk of the remaining lands.

Some of the items in the cruisers' report may interest present-day loggers:

250,000 Cedar Poles 30 feet and up at	$ .50
200,000 Cedar Ties at	.10
16,000 Cords Spruce Pulpwood at	1.50

White Pine (five logs per M) $15; Hemlock, $2.50; Birch, $3.00; Maple, $2.50. We cruised only large timber and paid no attention to hemlock and tamarack ties, balsam and hemlock pulpwood, hemlock bark (then used by tanneries), poplar, white birch or jack pine.

The Union Trust Company of Detroit, to which we sold $350,000 six percent bonds, and $175,000 First Preferred seven percent stock, sent an "expert" up to check our cruises, and he reported back "Our estimates overrun considerably . . . we find their estimates conservative and will on 'cutting out' show a good overrun." This was a master understatement. We knew very well that there was much more timber on the lands but we only went far enough to satisfy the Trust Company that their bonds were safe.[66]

At a Liquidation Sale in December 1925, the Consolidated Lumber Co., after lumbering for thirteen years, advertised for sale at public auction:

119,143,000 Feet of Saw Timber

33,377 Cords of Spruce Pulpwood

369,350 Cedar Ties

Timber grows faster than some folks think, but not quite that fast. Nonetheless, there was probably two or three times as much timber on the lands in 1912 as the amount stated in the Union Trust Co.'s bond circular.

We knew that any value we put on the mills, lumber yards and city property would be questioned in view of a dwindling timber supply, so we engaged the American Appraisal Company, and their valuation of $480,108.00, after deducting depreciation, was accepted without question by the bond houses.

It took us most of the spring and summer to secure the data and assemble it in shape to interest an investment house, a job which required a great deal of confidential book work, typing and correspondence. We imported Miss Ethel Kilpatrick, a young lady whom I had known from childhood and whose father had been my guardian until a year after I became head bookkeeper for the C. L. Co. She secured a leave of absence from a responsible position with the Pittsburgh Manager of the Philadelphia Company to come up and, ostensibly, work in our insurance agency, but in reality to be my confidential secretary in setting up our plans.

FINANCING THE DEAL

A snapshot of the financial background and money conditions prevailing in 1912 will explain how we were able to close a deal which would be impossible today. The federal government in those days took no notice of or interest in private business, except for the uniform reports required from railroads and public utilities by the Interstate Commerce Commission. There was no Securities Commission and no income, sales, social security, inheritance or unemployment taxes. The Federal Reserve System had not been organized and country banks kept their reserves with big banks in the Reserve Cities—New York, Chicago and others. Gold was in common circulation. Banks in towns like Manistique, paying three percent on savings deposits, where industries had passed the borrowing stage, had to go outside for loans or let the money lie idle.

There was no lack of venture capital. Businessmen and large depositors were looking for investments, but could stay out if they didn't like the venture. (There is even more venture capital today, but some of it is everybody's money ventured by government bureaus, and the private individual has no choice.)

This was before the era when country banks began to invest heavily in corporation bonds, with which they were more or less unfamiliar at that time. Note brokers, like A. G. Becker & Co. or W. T. Rickards, sold notes of Swift & Co., Marshall Field, Pillsbury and others in $5,000 and $10.000 denominations to country banks. For years the First National Bank of Manistique loaned $40,000 to $50,000 to wool growers, lumbermen and merchants in Union and Wallowa counties, Oregon, a very prosperous and growing country then, where there were more borrowers than lenders. We took only surplus lines from the Oregon banks and had them all checked by my brother, S. D. Crowe, who operated the telephone company serving those counties. We got eight percent on those loans, were never asked for renewals and we never lost a cent on them, but it would be illegal business today.

Michigan and Wisconsin pine men were investing heavily in west coast timber and timber bonds were popular with the big bond houses, who found them "sellers."

This was the background that enabled us to organize the Consolidated Lumber Co. and sell the Union Trust Company of Detroit a six percent timber bond issue of $350,000 and an issue of $175,000 First Preferred Stock, which, with an issue of $475,000 Second Preferred Stock sold to individuals with a substantial bonus of Common Stock, and a cash advance by A. Weston & Son of $120,000 without interest on four million feet of white pine to be delivered by August 1, 1913, and an advance of $8,000 by Edward Hines Lumber Co., completed our financing.

In fact, we found it easier to finance the proposition than to set up a satisfactory organization to run it. One factor that influenced the Union Trust to come in was that they had financed and were heavily invested in the Lake Superior Iron and Chemical Co., afterwards reorganized as the Charcoal Iron Co. of America, which had plants at Newberry, Manistique and near Marquette. The Chemical Co. and the C. L. Co. owned alternate sections like a checkerboard near Shingleton, which together formed a solid block of thousands of acres of virgin hardwood timber. The Trust Company was also involved with the M&LS Railroad from Shingleton to Manistique at that time, and I prepared maps and data and spent three hours one afternoon with Frank Blair, president of the Union Trust Company, showing him how those lands could be consolidated and the timber brought to Manistique over their own railroad to an enlarged plant at Manistique, the logical place for it. Mr. Blair seemed thoroughly sold and enthusiastic about the proposition, agreed to buy our securities and took me over to the law offices of Beaumont, Smith and Harris where he left orders for drawing up the papers.

The Union Trust Company, however, were better stock promoters than industrial managers, and for some reason I never could fathom spent large sums enlarging their Newberry plant and paid the DSS&A Railroad to freight their timber to that point, while M&LS Railroad lived up to its lumberjack name "Haywire" and was later sold for taxes. The Chemical Co.'s Manistique plant, the Charcoal Iron Co. of America, after a brief period of war prosperity, also went "haywire."

Lou Yalomstein and I had our own affairs to look after and neither of us wanted an active part in the management of the Consolidated Lumber Co. We tried to induce Charles Orr, Oren Quick and Norman Fox to come in and take the management, but the second generation was not interested, which was a pity because they would at least have avoided the costly mistakes made by Mr. Harmon and his Woods Superintendent, Fred Cooper. We were looking for a manager and Mr. L. C. Harmon came up from Menominee, where he had just sold out of some business enterprises, with $60,000 cash and good recommendations as an all-round businessman of broad experience. He seemed to be just the man for us, as the Consolidated Lumber Company was essentially a liquidating proposition and our overall plan was to sell the stores, hotel, water power, dwellings and vacant property; open up the town; get in new industries; and let the lumbering end of it work itself out. The C. L. Co. was a very efficient, harmonious, going organization with their camps and jobbers working, and the season's cut all planned and well under way with expert foremen and seasoned crews. The mill was in apple pie order ready to start on an hour's notice, and we had sold the lumber in advance at a price slightly above the market.

One of the strongest points we stressed when selling our securities was the smooth working organization, which we had no intention of disrupting.

The deal went through on schedule and the Consolidated Lumber Co. started out with Mr. L. C. Harmon as President and General Manager, I was Vice President and Lou Yalomstein was Secretary, neither of us being active at first. Mr. C. E. Kelso was Assistant Secretary and continued in charge of the office force, and W. T. Bradford of the Union Trust Company was Treasurer. Everything looked rosy.

CONSOLIDATED LUMBER COMPANY
Balance Sheet

ASSETS

		December 12, 1912
LAND 161,000 A at $3 per A. incl. timber on 135,500 A not cruised		$ 483,000
STUMPAGE Timber on 25,500 A as per independent cruiser's report		728,500
DEADHEADS Estimated value of sunken logs in rivers and lakes		25,000
PLANT ACCOUNT Sawmills, yards & equipment. per American Appraisal Co.'s figures		391,000
MISC. PERSONAL PROPERTY & EQUIPMENT		47,441.27
WATER POWER Engineers' estimate, 4,071 horsepower		150,000
CITY LOTS AND ACREAGE Vacant city lots and acreage approx. 90% of City's area not incl. yards and mills		130,000
CITY RENTAL PROPERTY Business bldgs. hotels, docks, warehouses, retail lumber plants, 147 two-story dwellings in good repair		318,377.57
CASH		27,500
		$2,300,818.84

LIABILITIES

FIRST MORTGAGE 6% BONDS Payable $50,000 per annum beginning Sept 1, 1913		$ 350,000
CAPITAL STOCK		
First Preferred	175,000	
Second Preferred	475,000	
Common Stock	750,000	
	1,400,000	1,400,000
SURPLUS AND RESERVES		
For organization expense	93,318.84	
For depreciation	427,500.00	
	550,818.84	550,818.84
		$2,300,818.84

COSTLY MISTAKES

It was unthinkable that anyone would dream of interfering with the woods operations in the middle of a logging season. We didn't know Mr. Harmon. As matters turned out, if we had all gone on a six months' vacation somewhere and let the old crews handle things on their own without interference we would have been about $200,000 better off.

I never could understand Mr. Harmon. I do not think the words "yes" or "no" were in his vocabulary, as he could make a complicated affair out of the most simple deal. To go from A to B he wouldn't go direct but would wander all through the alphabet and come back to B through C—providing he didn't get lost or sidetracked on the way. He convinced the Union Trust Company, which was influential on our Board of Directors, that the Chicago Company's ideas of lumbering were old-fogey and out of date, and for several months was given a pretty free hand as President and General Manager. He was very jealous of anyone connected with the old company. He was the modern businessman, who was going to show the old crowd how it should be done, although it developed that he didn't know a pine log from a hemlock.

The first costly mistake was when he and his Woods Superintendent, Fred Cooper, shut down—in midwinter—a couple of the Chicago Company camps that were cutting pine, and started a new camp in mixed timber. The reason given was that they wanted to "save some of the pine to 'sweeten' next year's cut," although Weston had contracted for the pine, and Hines would take all the hemlock and hardwood we could cut without any "sweetening." The Consolidated camps' logging costs that winter were at least $2 per M more then they should have been.

Harmon's next big mistake was to replace the C. L. Mill foreman and head band sawyer with men imported from Menominee, which created bad feeling and made the mill crew sore and sulky.

On top of that, by advice of the new foreman, although the mill was all set and ready to go, he installed a new re-saw and made other changes costing over $30,000 before they had ever seen a wheel turn. The changes were all for the worse because the new machinery had a lot of "bugs," spoiled a lot of clear pine lumber, reduced the daily cut and delayed the starting of the mill nearly thirty days in the spring of 1913. So on August 1st, when our contract with Weston called for four million feet to be in pile, we had less than one million feet, and we were forced to revise Weston's contract, pay him interest from the previous December and carry his lumber over to the next shipping season. The extra interest, taxes and insurance on Weston's lumber cost us $16,768.

Things eventually got so bad that even the Trust Company woke up. Mr. Bradford came up from Detroit. An emergency meeting of the directors was held. Mr. Harmon was let out and the job was shoved onto me, with instructions to do the best I could to pull the company out of the hole. I accepted, providing I be given an absolutely free hand and that the Union Trust Company would cooperate. Mr. Bradford agreed, saying, "It's your baby now, but we'll help you nurse it." When I was catapulted into the management of the Consolidated Lumber Company in 1913 it was a pretty sick baby. The National Bank of Commerce in Detroit had just protested a $15,000 note; we had lost the goodwill and respect of the old crowd and our customers. We owed $273,926.34 to banks and on open accounts, much of it past due. Besides that, we owed Weston and Edward Hines

$180,000 for cash advances on lumber that was not being delivered according to contract either in quantity or quality. In fact, on August 1st we had only about twenty-five percent of the lumber called for in pile; the mill crews were nervous and edgy and had no confidence in the new foreman. Accidents and breakdowns were frequent, the mill was cutting only seventy percent of normal capacity, and the first installment—$50,000—of our bonds was due in less than ninety days. The general opinion was that receivership was inevitable and only a couple of weeks off.

It was late in the evening when the emergency meeting of the directors closed. We arranged with Mr. Joy, President of the National Bank of Commerce, to extend the protested note for thirty days. The next morning I went to Mr. M. H. Quick and said. "Mr. Quick, I don't have to tell you about the fix we're in. I wish you'd go down to the mill with me and tell me exactly what you would do, in this situation." He said, "Will, I'll glad to tell you what I would do, and we don't have to go down to the mill. I would stop using that new horizontal re-saw, which is degrading a lot of your good pine lumber; put the vertical re-saw (which had fortunately been left in position) back in commission; give Drell Carr his old job as band sawyer; and I would put George Gayer in as mill foreman. That man Russell doesn't understand the mill or the timber and the men don't like him."

I did exactly that. The atmosphere cleared immediately. The men caught the spirit, and in four days the mill was averaging 215,000 feet a day instead of 148,000 and there was no more too thick or too thin lumber.

I also let Fred Cooper go at once and engaged Frank Cookson, giving him full and complete authority as General Woods Superintendent, and from that time on there was no more foolishness in that department.

After making these changes I took George Gayer and got Charley Hall, one of Hines' field men, to go to Chicago with us to see Edward Hines, who was on the warpath. When we explained the changes we had made and what we were doing, he not only extended our contract but gave us an additional advance and invited us out to Sunday dinner at his home in Evanston.

The weather that fall was a great break for us because we were able to run the mill up until after Christmas, over a month later than it had ever been run before, with the result that we caught up on Weston's contract, which greatly pleased them.

Concurrent with this, we carried on a campaign to sell the Ossawinamakee Hotel, the stores, houses, cut-over lands and other property not essential to the lumber business, with fairly good success.[67]

The intake pipe leading from the dam at Sart's on the Indian River to the city water reservoir needed replacement. We made surveys and maps and laid out a tract (including fifteen acres donated by the Women's Club) of approximately sixty-five acres as Riverside Park at the present intake and sold it to the city, together with 410 HP potential water power in 1914 for $20,000. This shortened the distance and saved the city the expense of building over a mile of intake.

Mr. E. H. Coyle of Greif Bros. Cooperage Company, Cleveland, Ohio, a reliable firm of excellent standing, came into our office with P. N. Snook of the Edwin Bell Company; they were operating the cooperage plant at Engadine. They were looking for a new site and timber supply. I showed them maps and estimates of available hardwood near Manistique and gave them a satisfactory price on a site on Houghton Avenue accessible to both railroads. They asked if local people would take a $25,000 bond issue. I asked them how much they would put in themselves. "About $50,000," they said. I called the local banks and a couple of businessmen, who agreed to take the bonds, and the deal was closed the same afternoon.

We sold the C. L. Store stock of merchandise to Lou Yalomstein, who organized the Peoples Store Company and built the Peoples Store building at Oak and Cedar streets.

In 1913 when I assumed the unwelcome job of Manager of the Consolidated Lumber Company, its Second Preferred Stock ($100 per share) was offered at $15 per share with no takers, and the Company was only one jump ahead of receivership. By March 1915, we had paid off $97,000 of the bonds, reducing the original issue of $350,000 to $253,000, and had net current assets of $140,881.60. At that time the company was sold to the Stearns interests of Ludington, who paid $10 per share for the Common Stock, which had originally been issued as a bonus to the Preferred Stockholders and which cost them nothing.

I never claimed any particular credit for this showing. All I did was to put the old C. L. Co. working organization back in charge and they worked it out.

When Stearns and W. T. Culver took over in 1915, the cream, the big pine, had all been skimmed off and it was from then on a hemlock and hardwood operation, but under a vastly different and more difficult set of conditions than in these days of chain saws, bulldozers, paved roads and modern logging trucks.

Stearns and Culver operated with varying success until they leased and then sold the mill and yards to the Stacks in 1922. Bruce Odell was their General Manager and they gave Harmon, who kept his stock and stayed with them, a nominal position without any real authority. They sold the water power to the *Minneapolis Tribune* (Murphy)[68] for around $75,000, but coincidentally contracted to deliver such a large amount of spruce pulpwood to the paper mill at such a low price that it absorbed all their profit on that deal. They sold the balance of the city real estate to Benjamin Gero, and finally sold their remaining assets at a liquidation sale in 1926.

The Stacks operated the mill for several years until they finally dismantled and sold it, and so the story of the lumberjack and big sawmill era in Manistique comes to an end.

The last of the C. L. Co.'s big pine was cut by the Consolidated Lumber Company in the winter of 1914-15 at Camp 85 from the SE Quarter of the SE Quarter of Section 10, in Township 44 North Range 18 West about ten miles northwest of Steuben, and the stumpage value at that time was over $1,200 per acre.[69]

Superior View Studio, Marquette

When the pine was gone: A view of cutover lands on the Kingston Plains, Alger County, in 1900. On a similar photo, Crowe wrote, "This land should be reforested."

CHAPTER EIGHT

Peace, Progress and Prosperity

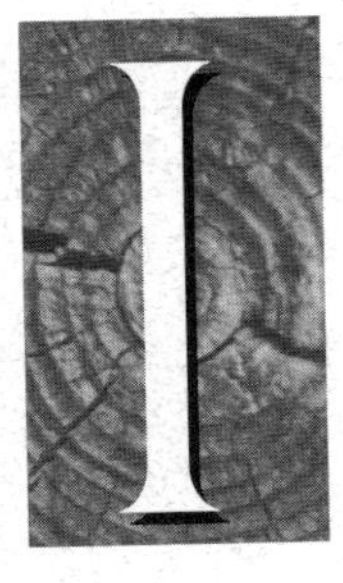

I was a firsthand observer of the people, events and settings which made up the peak years of white pine lumbering in the Upper Peninsula of Michigan, from my vantage point with the Chicago Lumbering Company of Michigan and the Weston Lumber Company, which built and dominated the City of Manistique for half a century. The general economic conditions, background, ways of life and habits of thinking prevailing in the colorful white pine lumber era made inevitable the destruction of the magnificent virgin pine forests which once covered the state of Michigan.

I will consider my time well spent if these chapters have helped to correct the general—though false—impression created by sensational writers that the old-time lumberjack was, as a rule, a rough and tumble "hell-raising timber beast" with only women and whiskey on his mind. I also want to correct the equally false and malicious stories that the big lumbermen were "robber barons" with a "public be damned" attitude who "stole the forests" from the government and "exploited" their employees and the public alike, whose only thought was to "Cut out and get out" and "to hell with the Country" and that they alone were entirely responsible for leaving a temporarily ruined fire-scarred country of pine barrens, charred stumps and timber slashings.

> "The saga of the old-time lumberjack in the woods and the whitewater man on the rivers . . . lasted almost one hundred years."

The economic history of the United States covers three fairly well-defined Epochs:

First Epoch: 1786 to 1860. Individualistic and Agricultural. People of mixed races from all parts of the world, free from government domination, lived together and built the foundation for greater progress in living standards in one century than had been accomplished by mankind in six thousand years of recorded history.

Second Epoch: 1860 to 1914. Individualistic and Industrial. From the Civil War to World War I. Although agriculture was still dominant, this epoch witnessed the most astounding development of mechanical invention, industrial organization by business corporations, mass production and a greater change and improvement in living standards and human ideals and welfare than any other period of history before or since.

Third Epoch: 1915 to date.[70] Reactionary. Emergence of government controls, submergence of individualism and centralization of political power at an accelerating rate. Expansion of military power. Growing ideas and dreams of world domination. "Union Now" and "One World." Increase in population, wealth, material progress, inflation of money. Professional sports, games and spectacles. Political promises to provide a more abundant life for everybody. Nature's parallel—the rapids above Niagara. History's parallel—The Roman Empire, bread and circuses.

This story has dealt entirely with the Second Epoch, because the story of the Chicago Lumbering Company of Michigan and the Michigan pine lumber era lies within and, in fact, coincides almost to a year with that colorful epoch of our history: Fifty years of peace and progress, barring a few months' interlude in the Spanish War skirmish.

The Chicago Lumbering Company, organized in the summer of 1863, ceased business almost coincidental with the outbreak of World War I in 1914. And the saga of the Michigan pine industry falls almost completely within that same period.

An idea has been fostered in some recent literature by writers who seem to lack a sense of perspective and proportion, and I think is generally entertained by the present teenage generation, that life in the United States prior to 1932, sometimes referred to as the Old Deal, was a static affair, an outdated horse-and-buggy age in which time stood still.

Nothing could be further from the truth. Old Deal times of 1910 were vastly different from Old Deal times of 1860, and the fifty years between Appomattox and Sarajevo were a boiling, seething, exciting stream of material progress unequaled by any other fifty-year period of all time. The so-called Gay Nineties were right in the middle of it.

A few of the colorful actors on the world stage in that individualistic age were: John Philip Sousa, Gilbert & Sullivan, Maude Adams, Ethel Barrymore, Sarah Bernhardt, John Drew, Joseph Jefferson, Schumann-Heinck, Melba, Patti, Caruso, Paderewski, Longfellow, Whittier, Mark Twain, P. T. Barnum, Buffalo Bill, John L. Sullivan, Beecher, Spurgeon, Moody & Sankey, Carrie Nation, the Vanderbilts, John D. Rockefeller, Hetty Green, Andrew Carnegie, Morgan, Hill and Harriman, Henry Ford, Thomas Edison, Marshall Field, Horace Greeley, Charles Dana, James Gordon Bennett. Life couldn't stand still or be dull with a cast like that.

In that short period came the telephone, electric light, typewriter, linotype, phonograph, bicycle, tractor, Bessemer steel, transcontinental railroads, refrigerator cars, sleeping cars, automobile, airplane, the Atlantic cable, steel steamships, oil fields, repeating rifles, radio, moving pictures, indoor bathrooms and the Soo Canal.[71] The reaper, mowing machine, sewing machine and cotton gin, then but recently invented, were making their influence felt in a large way.

Take away these things—all invented, not by government planners, but by old fogies of the horse-and-buggy days—and our modern way of life would be impossible. Our big cities—New York, Chicago, Detroit—would starve in a week while crops might rot in Iowa, Kansas, Nebraska and Texas.

And observe: These wonderful inventions, which had such a tremendous impact and influence on living standards and habits, were all constructive and not destructive. It was an age of peace and not war. The glory of war and conquest had been thoroughly debunked by our own Civil War. We were full of boundless optimism and dreams of a wonderful future. There was a song in our hearts and everybody was saying, 'What next?" When a new invention came out, our first thought was not how it could be used to kill other people, or as a defense weapon. With only one-third of the population we have now, we weren't afraid of any other nation or combination of nations on earth, and no other nation would have dreamed of attacking us.

During this fifty-year period we were on the friendliest of terms with every nation on earth. Russia was always friendly to us. It was fear of Russia and the influence of Queen Victoria and Gladstone that kept England from jumping into the Civil War on the side of the South because we had put an embargo on cotton exports that England needed for her cotton mills.

China was very friendly—thanks to traders like Robert Dollar and his round-the-world steamship line. Our fast clipper ships were trading on every continent, bringing wheat from Australia to England and so on.

(And our so-called present day liberals and New Dealers have the gall to say that we were Isolationists!)

KINGDOM OF WOOD

Wood, not steel, was the predominant building material in this epoch. It was the great era of railroad building and western settlement, of large families, frame houses and barns, large wooden buildings like the present three-story Ossawinamakee Hotel, C. L. Hardware and Park Hotel (the former C. L. Store and Office). The large Hiawatha Hotel that stood near the Soo Line depot; the American House and St. James hotels where the Sinclair Oil stations now stand; the Central, Westside and Lakeside schoolhouses; the immense sawmills and many business blocks in Manistique are typical of frame buildings in every city and town. Chicago itself was built largely of wood at the time of the great fire in 1871. Board fences, sidewalks, bridges, railroad trestles, railroad cars themselves—all wood.[72] Charcoal iron furnaces in Fayette, Manistique, Gladstone, Marquette, Newberry and other Michigan towns consumed immense quantities of hardwood in their kilns. There were even wooden highways. The plank road from Butler, Pennsylvania, to Pittsburgh ran within three miles of our family's farm. Streets were often paved with wood blocks. Ludington Street in Escanaba was paved with round cedar blocks on end. The big sawmills in the Manistique area each operated a lath and picket mill in connection with the sawmill and the White Marble Lime Co. had a large shingle mill. Try to buy a bundle of wooden lath or shingles today.[73]

Immigrants by the million were pouring into the country to become good American citizens regardless of how they had fought each other in Europe; most of them settled in the great West. Instead of displacing American labor, they created new markets and jobs and inventions.

And that's the way it was. On the one hand a whole state covered with the finest virgin pine timber, easily accessible by cheap water transportation. On the other hand, settlers by the million pushing into the great prairie states, and railroads pushing across this inland empire, all demanding billions of feet of lumber and forest products. That wonderful prairie empire, the world's greatest bread basket, could never have been developed without the Michigan-Wisconsin-Minnesota forests.[74]

You who are so ready to condemn the Robber Barons for wasting our natural resources, answer me this. Who needed that pine timber most: The railroads and settlers who were building the western empire in that wooden age? Or we of today's steel, concrete and plastic age?

And tell me this. Just exactly what would we do with six or seven hundred billion feet of virgin pine timber if we had it today?

Timing is just as important in a nation's economy as it is in the engine of your automobile. The *National Geographic* magazine, which I am sure no one will accuse of distorting facts, has this to say on page 304 of its March 1952 issue: "Michigan, with its pine lumber, largely built the Prairie States. In the 1870s it led the nation in production; its 1,600 sawmills cut more than three billion board feet a year."[75, 76] The government was selling its natural resources for a song and giving away public lands in million-acre hunks to railroad builders, and it has been criticized for doing that by folks who measure everything with today's yardstick. But, what is land worth without people? In the beginning the government owned all the land, timber and minerals, but they were utterly without people, population and settlers. The government wasn't taking things away from people then. It was giving things to people, and the opportunities were open to all. It was simply trading land and idle resources for free settlers and populations, and that is what made this country so great. If the government had kept all these resources as some think it should have done, we would have an economy exactly like Russia has today, peopled with a subject race.

The big lumbermen were not fools. They knew that timber values would increase with growing population and many of them did not want to cut out and get out, but were forced out and taxed out. A prominent member of the Schoolcraft County Board of Supervisors told me his idea was "soak the C. L. with all the taxes we can before they cut timber, and to tax their logs in the river for all they would stand." I told him, "You won't gain anything, because you will only force them to cut it faster."

In those days there were no sales or income taxes; local and state governments were supported almost by direct taxes on real estate and personal property whether the property was earning an income or not. Under conditions like that, a lumberman, no matter how many millions he had, would go broke if he tried any long-term forestry program. He had absolutely no other choice except to cut out and get out.

The saga of the old-time lumberjack in the woods and the whitewater man on the rivers is practically over. It lasted almost one hundred years. Beginning with the rising sun of the coast of Maine, it moved steadily westward to the shores of the western ocean on the rugged coast of Oregon. The Penobscot and the Androscoggin, the Susquehanna and the Allegheny, the Ottawa, Muskegon, Saginaw, Manistee, Au Sable, Tahquamenon, Manistique, Menominee, Brule, Wisconsin, the St. Croix and the Mississippi, Old Man River himself, all in turn echoed to his shouts and songs, and witnessed his feats of heroism and daring.

Then came a long trek across the great plains and mountains to a country where he was met and challenged by forest giants beyond anything he had ever imagined, which compelled him to exchange his hand tools and individual methods for giant machines and engineering operations on a huge scale. He has met the challenge, and can go no further; he has dug in and fortified himself with tree farms, selective cutting and protective methods against fire and other hazards. He proposes to go on cutting trees forever.

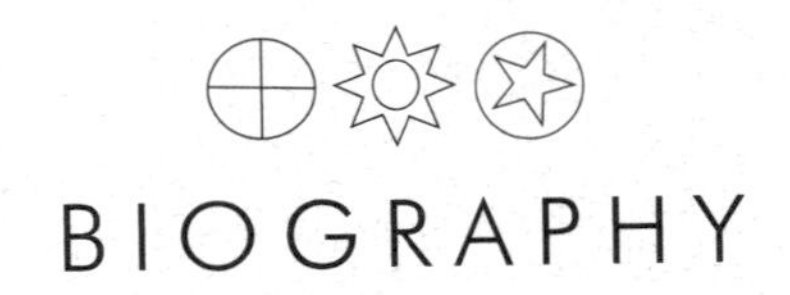

BIOGRAPHY

William Scott Crowe
1875~1965

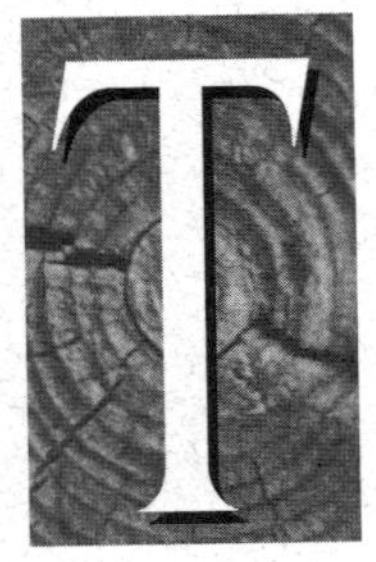

The young man stepped off the deck of the steamer *City of Ludington* into, as he later wrote, "a strange new world such as I had never seen or dreamed of." It was exactly midnight on May 29, 1893, in the port of Manistique, Michigan, at the head of Lake Michigan. Before this trip, which started in Pennsylvania, he had never seen a ship, a large body of water, a sawmill or even a big tree. He knew no one in this small town in the Upper Peninsula of Michigan.

He was only seventeen years old, yet in the past four years he had already lived through a dizzying series of events. He could not have imagined the turns his life would take from this night in 1893, centering in this small town and in the lumbering world of Michigan's northern lands.

William Scott Crowe, the oldest of six children, was born in New Castle, Pennsylvania, on September 23, 1875. His parents, Mary Amelia (Millie) White and Oliver Cameron Crowe, were married in 1874 in New Castle and settled on a family farm in Butler County, twenty miles north of Pittsburgh. The Crowe family was quite well known in eastern Pennsylvania; an uncle, the Reverend Samuel J. Crowe, was the first president of Geneva College in Beaver Falls, Pennsylvania. After William (known as Will), came Maria Mae, who preferred to be called Mae; Samuel Dean, always known as Dean; Margaret Emma; and Milo Cameron (born on Will's 11th birthday). Lulu, the youngest child, was only an infant in 1889 when both parents contracted tuberculosis, then known as "consumption." The medical advice of the day was to find a drier, warmer climate and hope for

Hendricks & Co., Allegheny, PA; Ellen McFeeters Turnbull collection

The Crowe family children in 1889, before the move to Colorado. From left: William Scott, Margaret Emma, Samuel Dean. Seated in front: Milo Cameron, Maria Mae holding baby Lulu.

the best. The Crowe family moved to the high plains of Colorado, near Greeley, but both parents died within six months of the move.

The children were placed in homes in Colorado where Pennsylvania relatives hoped the climate would help them resist the infection; in fact, none of the children did contract tuberculosis. While the younger children were cared for by local families in and around Evans, Colorado, Will went to live, at age fourteen, "on a cattle ranch on the treeless plains of eastern Colorado northeast of Fort Lupton." He and Mae attended school nearby. Dean and Margaret were taken in by the Morrison family, who later adopted Margaret; baby Lulu was adopted by a well-to-do couple without children. Milo described his new home with a pioneer couple as "on the open prairie in a partial dug-out house." Will saw his brothers and sisters only occasionally for the next two years. In 1891, uncles found an adoptive home in Jamestown, Pennsylvania, for Milo, who was not quite five years old. Will and Mae took the little boy back east by a three-day train trip (away from what he later remembered as fifteen months of a "half-wild but happy life in the Colorado dugout"). Milo wrote later that the McFeeters had really wanted to adopt his baby sister Lulu, but she was already taken so they took Milo "on trial" for a year or so. At first, and for some time afterward, Milo was known as "the Crowe-boy." Dean, Margaret and Lulu stayed at their homes in Colorado. Dean later ran away from the Morrisons in Colorado where he had been chore boy, doing the milking for two or three years.

With the help of his guardian, the Reverend A. Kilpatrick, of Valencia, Pennsylvania, Will enrolled in nearby Grove City College for a nine-month business course. He learned later that the Reverend Kilpatrick had asked Dr. Ketler, president of the college, to look after the young man.

As Will Crowe remembered it, "After graduating (in 1893), I was staying at an aunt's house in Mercer, Pennsylvania, all packed up and ready to leave for Colorado the next day, when I got a letter from the Chicago Lumbering Company of Michigan, sent by J. D. Mersereau, offering me a job in their office at Manistique, Michigan, at $40 per month and telling me how to get there. My, how important I did feel. I wondered for a long time how Mr. Mersereau had ever heard of me because I had never heard of him or knew that such a place as Manistique existed.

S. M. Gillespie, New Castle, PA; W. S. Crowe collection

Oliver Cameron Crowe and Mary Amelia White, parents of William S. Crowe, were married in Pennsylvania in 1874.

William S. Crowe, circa 1898, in his early twenties.

H. I. Anderson, photographer, Manistique, MI; W. S. Crowe collection

"It seems that Mr. N. P. Wheeler had met Dr. Ketler and mentioned that Mr. Mersereau was looking for an office assistant, and Dr. Ketler mentioned me. On such chances, the whole current of our lives sometimes changes." [3]

Some weeks later, the *City of Ludington* delivered him to the Manistique docks. It was the peak of the white pine lumbering era in the Upper Peninsula of Michigan.

Almost sixty years later, the memories that Will Crowe recounted of that night and the scenes as he entered his new life were vivid.

On the first Sunday, he and a companion walked three miles out to Indian Lake, through groves of virgin white pines that he later described as "so tall they blotted out the sun and no undergrowth grew beneath them." He looked out across the lake, eight miles long and three miles across, and saw it filled almost from bank to bank with white pine logs, rafted up for the run down the Indian River to the mills. They walked halfway across the lake on the logs. Years later, that first excursion through the forest still formed an indelible picture in Will Crowe's mind. He spoke of the white pines standing so thickly that sunlight could not reach the forest floor; there was no undergrowth and the first branches of the trees started sixty to eighty feet above the ground. He remembered the intense quiet in the groves except for the sound of wind high in the tops of the giant pines.

His life in the Upper Peninsula of Michigan had begun. It would not end for over seventy years, a rise through the lumbering business, a central place in his adopted town and the gathering of his experiences in a book called *Lumberjack*.

Will, as the oldest of the Crowe children, had promised his parents that he would keep in touch with all of his brothers and sisters, wherever they were, and he fulfilled his pledge. He was the family correspondent and arranged visits both in Manistique and later at their homes across the country. Several months after Dean had hopped the train away from the Morrisons' farm in Colorado, Will found out where he was and had him come to work in Manistique. Dean spent several other periods of his personal and business life with Will, and they entered into several business ventures together. Milo worked as a teller in the First National Bank in Manistique during one summer, and remembered having to work after

[3] WSC letter to Mr. Timothy N. Pfeiffer, February 2, 1953

hours for three days before he found a 6-cent imbalance in his accounts. Margaret Morrison Hesslink, after living in California, came to stay with Will in her later years.

From the first position as time boy and office assistant, Will Crowe was soon promoted to head bookkeeper, a post he held for seven years (1893-1900). In 1900, he and other Manistique businessmen organized the First National Bank; he held the position of cashier and later president for thirty-four years.

In 1912, Crowe and Lou Yalomstein, manager of the C. L. store, were offered the opportunity to purchase the business and holdings of the Chicago Lumbering Company. The sometimes rocky course of that lumbering business endeavor is detailed in Chapter Seven.

After he had lived in Manistique for seven months, Will was introduced to Bertha May Orr at a Baptist church skating party. Will, formerly a faithful attendant at the Presbyterian Church, found a keen interest in the doings of the Baptist Church from then on. He later delighted in retelling the tale of how he won the hand of Bertha May away from several other suitors. Will and Bertha were married over seven years later, on June 5, 1901. Bertha was the daughter of Burton Orr, one of seven Orr brothers who settled in Manistique in the early 1870s from New York State. Like Will, Bertha was the oldest child of the family, left with responsibility for the younger children at age fifteen when her mother died. Her brother, Wesley Burton Orr, was only two years old at the time of their mother's death. Will and Bertha's home on Lake Street in Manistique became Wesley's home, along with their own children: one son, Stanley Oliver, and two daughters, Helen Elizabeth and Ethel Mae.

The children had fond memories of the house at 111 Lake Street, with the gardens, playhouse and backyard cliffs where Will had a long toboggan slide built for winter fun. The run-out from the slide extended across the road (no automobiles in the wintertime) and into the courthouse grounds. When Ethel broke her leg on the slide one winter, down it came!

Will and his brother-in-law, Wes Orr, later purchased land on Indian Lake, about three miles from town and built Sunset Beach

Butts Studio, Buffalo, N.Y.; W. S. Crowe collection

William S. Crowe and Bertha May Orr Crowe, June 1901, on their wedding trip.

William S. Crowe, lumberman, banker, insurance agent and real estate broker, at age 83.

W. S. Crowe collection

Resort, with rustic cabins and a "tearoom" where Sunday dinners were served to area residents and tourists. Family cabins were built and several are still occupied each summer by Orr descendants.

In 1934, Crowe retired from the bank; he continued in his real estate and insurance businesses. He was a Hartford Insurance Company representative for over fifty years, and acquired extensive lake and forest lands around the area. One of his holdings on Indian Lake was later purchased by the State of Michigan for the west unit of Indian Lake State Park.

With little opportunity for formal education, Will Crowe studied history, banking, economics, English, gardening and other subjects and acquired a large library of books on many subjects. In the early 1900s, he was the pitcher for the Manistique indoor baseball team and participated in bicycle races in the Upper Peninsula. He loved to travel and took his family on three separate trips to the West Coast during the 1920s. Bertha Orr Crowe died in Manistique in 1946; his son Stanley and daughter Ethel also preceded their father in death. Will and Bertha had three granddaughters, Elizabeth Ann McGlothlin, Helen (Lynn) Crowe McGlothlin and Phoebe Jane Giffen, and two grandsons, John Crowe Giffen and Robert Bruce Giffen. Will Crowe was able to meet some of the next generation when great-grandchildren visited at the family home in Manistique and the cabins on Indian Lake. At age eighty-five, he traveled to Europe and then visited England, where he suffered a broken leg, flying home while still in the cast.

William Crowe was active in correspondence, reading, writing, real estate and other business deals and chess contests by mail until he was hospitalized six weeks before his death at age ninety on October 29, 1965.

Left behind on his office door in his precise "Palmer-method" script was a small notice:

Back at 1:00 p.m.

SOURCES FOR BIOGRAPHICAL INFORMATION

William S. Crowe. *Lumberjack, First Edition, 1952*

Helen C. McGlothlin. *Foreword to Lumberjack, Second Edition, 1977*

William S. Crowe. *Correspondence and memoranda*

Milo McFeeters. *"Letters to the Children"*

Ellen McFeeters Turnbull. *"The McFeeters Story" and other materials*

Editors' Notes

CHAPTER ONE

1. By an 1842 Michigan law, log marks (at first only bark marks) were to be registered at the clerk's office of the county in which the logs were floated and milled. In 1859, local legislation required log owners on the Muskegon River and its tributaries to stamp their marks in several places on the ends of the logs. Later, state legislation required marking and registering throughout Michigan. Over three hundred log marks were registered in Houghton County. Hesterberg, Gene A., and U. J. Noblet (Forestry Department, Michigan College of Mining and Technology). "Annual Spring Drives on Copper Country's Mighty Sturgeon Recalled." *Marquette Mining Journal* (June 15, 1957).

The fee to register each log mark was 25¢. A single company might have many log marks. Delta County had so many log marks that a Log Mark Record sheet was developed. Charles Hebard of Pequaming also had dozens of marks; they were registered in Baraga County throughout the 1880s and 1890s. Rod Smith, Education Consultant, retired, Michigan Department of Natural Resources Upper Peninsula Region, Marquette, MI, conversation with Emerick, January 2002.

A hand drawing of the many Pequaming log marks is displayed in the Logging Museum of Newberry. "The Diamond (Match Company of Ontonagon) possessed more than 150 registered log marks. There were scores of diamond marks with letters. . . . there were bulls' heads and watch marks, snowshoe and foot marks and bar circles galore." "Ontonagon Keeps Great Reputation for Timber Stands." *Marquette Mining Journal* (September 2, 1958).

Bark marks, also called hack marks, were made with ax strokes in several different places about three feet from the end of the log, after a slab of bark was removed. Bark marks might indicate owner's name, property where the tree was cut, season it was cut, camp number or other identifying information. They were used only until about 1910. Since there were a limited number of ways to cut ax marks, log marks were later used to expand the number of variety of ownership marks.

Log marks were pounded into the end of the logs with a metal branding hammer; the marks were also called stamp marks or water marks. The hammer had to be swung hard to make a deep imprint in the green log. Because of the compression of the wood fibers back into the log, the mark was still visible even if the end of the log was mutilated or cut off.

Not all log marks were registered as required by law. Often the larger companies (note: including the Chicago and Weston lumber companies) which controlled the logs from woods to mill did not register their marks, but they were known and recognized. Sources for information on log marks and bark marks: Smith, interview; Bacig, Tom, and Fred Thompson. *Tall Timber—A Pictorial History of Logging in the Upper Midwest*. Bloomington, MN: Voyageur Press, 1982; Bachman, Elizabeth M. "Log Marks, Brails and Booms." *The Naturalist*, v. 11, no. 4, Winter 1960; Works Progress Administration, compiler. *Michigan Log Marks*. East Lansing: Michigan Agricultural Experiment Station, 1941; Hargreaves, Irene, and Harold Foehl. *Logging from A-Z*. Bay City, MI: Red Keg Press, 1988.

2. Rod Smith of Marquette shared with the editors (January 2002) his extensive collection of log marks, researched from the county records throughout the Upper Peninsula. Well over one hundred years have passed since the lumberjacks marked the mammoth white pines to be driven down the many rivers of the peninsula—the Driggs, Fox, Tahquamenon, Two-Hearted, Dead, Escanaba, Manistique, Ontonagon, Fence, Indian, Sturgeon and many more—to mills on the Great Lakes. Those years melted away and words on pages became real when Smith brought out an end piece from a log found in the Driggs River near the present Highway M-28. There, stamped clearly in several places, was the Barred O mark (⊕) described by Crowe.

3. Existing railroad companies were reluctant to allow logging railroads to cross the main lines; the purpose of the purchase of Hall & Buell's existing railroad line was to give the Chicago Lumbering Company access across the Soo Line Railroad right-of-way.

4. Pequaming, on Lake Superior in Baraga County, was another example of an almost completely "company town." Charles Hebard's company built and owned all of the buildings, including the business places. The company eventually built and rented about one hundred homes for its workers. The company, including its timber holdings, the town, railroad and mills was purchased by Henry Ford in 1923. Cleven, Brian. "Henry Ford's Tasty Little Town—Life and Logging in Pequaming." *Michigan History* (Jan-Feb 1999): 19-23.

5. "Manistique was a bustling log terminus when New York timber company owners Abijah Weston and Alanson Fox bought the then-struggling C. L. Co . . . the pair created a local timber empire. . . . Initially pine logs, straightest and tallest, were valued for ship masts and hauled out of the woods without being cut. But in the years just after the Civil War, Weston and Fox sought to take advantage of the new postwar demand for wood to build homes, other structures." Pepin, John. "Area's Logging History Recalled." *Marquette Mining Journal* (March 25, 2001): 1A,11A.

6. The Ossawinamakee Hotel, most often called the Ossa, took its name from the last chief of the Chippewa Indians in the local area. *Manistique Centennial Book*. Manistique, MI: Manistique Centennial, Inc., July 1960. In 1901, "the following good hotels cater to the wants of the strangers within our gates: Ossawinamakee, Hiawatha, Hotel Barnes, Tretchler Hotel, Keystone and St. James." *A Souvenir of Manistique, Michigan*. Manistique, MI: *Manistique Härold*, 1902; reprinted by Schoolcraft County Historical Society, 1992, not paged.

7. Fred Orr was elected to two terms as sheriff and served from 1915-1918—Schoolcraft County (MI) Sheriff Gary Maddox, conversation with Emerick, January 2002.

8. Indian Lake was considered "still water" and the C. L. Company paddle-wheeler was needed to raft the logs into booms and to tow them across the large lake to the outlet into the Indian River so they could be floated to the mills in Manistique.

9. "Haywire" was a lumberjack nickname for the Manistique and Lake Superior Railroad, which started in 1909 as a successor to the Manistique and Northwestern Railroad. The M&LS line ran from Manistique to Shingleton with a stop in Steuben for those passengers who wanted to have lunch at George Hughson's hotel. Mrs. Hughson had a reputation for the best food around and it was "all you can eat" for 50¢. The line ran two times per day for passengers and freight. Groceries were hauled in via the "Haywire," unloaded at the post office/hotel and hauled by horses to the logging camps in the area. Most of the big logging camps along the line stopped operating in 1920, and the railroad eventually went bankrupt. In 1970, the railroad right-of-way was acquired by the Michigan Department of Natural Resources and the U.S. Forest Service for a snowmobile trail corridor. Lesica, Ferdinand. "History of Schoolcraft County." manuscript, n.d. (Lesica was Schoolcraft County clerk from 1957-1984.); Alger County Historical Society. *Alger County—A Centennial History 1885-1985*. Munising, MI: 1986; Orr, Jack. "Memories and Myths." *Manistique Pioneer-Tribune* (February 1, 1979).

"Does anyone wonder how the M&LS Railroad acquired the name 'Haywire"? . . . the motive power for logging along the railroad was horsepower. . . . they required baled hay. After a time there were small mounds of haywire from the baled hay, all along the tracks. . . . The engineer would head for the nearest pile of haywire and try to make emergency repairs (on the engine and railcars) good enough to get back to the roundhouse and shops." Orr, Jack, quoting Vern Noble. *Lumberjacks and Other Stories*. Manistique, MI: Privately published, 1983, p. 48.

10. Onota, a settlement west of the current town of Munising, was actually the first county seat of Schoolcraft County, selected in 1871; after a disastrous fire in Onota in 1877, the county seat was moved to Manistique in 1879 and set up offices in the main office of the Chicago Lumbering Company and the Union Hall. In 1881, the Chicago Lumbering Company deeded the entire Block Number 8 of the village of Manistique for the courthouse, jail and sheriff's quarters. *Alger County—A Centennial History 1885-1985*. 1986; Lesica. *History of Schoolcraft County*.

11. John Munro Longyear was a famed landlooker in the Upper Peninsula, beginning his work in 1873 out of Marquette on Lake Superior. He spent five years in the rugged backcountry, surveying timber lands, later becoming land agent for the Keweenaw Canal Company and owner of iron mining interests. His landlooking experiences are detailed in Helen Longyear Paul's book, *Landlooker in the Upper Peninsula of Michigan*. From the reminiscences of John Munro Longyear. Marquette, MI: Marquette County Historical Society, 1960.

In a note on a letter dated September 14, 1953, from the firm of Milbank, Tweed, Hope and

Hadley of New York, Crowe wrote: "John M. Longyear made a fortune in Upper Peninsula timber and mineral lands. He built a large stone house in Marquette—almost a castle—on a hill overlooking Lake Superior and surrounded it with a big stone wall. He was peeved at the community when the City Fathers allowed the LS&I Ry to build its tracks along the waterfront in front of his residence, so he took the building down stone by stone and shipped it to Boston."

12. William Burt was another noted Upper Peninsula surveyor, who ran lines in many locations, including Marquette County. In the fall of 1884, Burt and his party were in charge of survey parties under Douglass Houghton.

Waring, Betty. *Yellow Dog Tales and Logging Trails to Big Bay, Michigan.* Big Bay. Privately published, 1986.

13. The spring is near the site of the former Arrowhead Inn, on the northeast shore of Indian Lake. Evidence of an Indian settlement and mission church, founded by Father Frederic Baraga, was reported in 1840-1850 on government surveys of the area. The surveys noted the Chippewa settlement was "at the outlet of Indian Lake, near springs." *Manistique Centennial Book*. 1960, p. 9.

14. George Hovey's name was given to Hovey Lake in what is now Alger County, north of the Chicago Lumbering Company's camp at Doe Lake. Hovey was camp foreman of Camp 47 at Hovey Lake starting in 1892; he also had charge of the upper log drive in the area of Camp 76, near the present Chain of Lakes. The Indian River, one of the key rivers for moving logs from woods to mills in Manistique, begins its course at Hovey Lake. *Alger County—A Centennial History 1885-1985*. 1986.

15. The origin of the expression "tunket" is obscure, but its meaning is vivid: A euphemism for "hell," usually as in what or where "in tunket." References of usage during George Orr's time include: *Scribner's Monthly* (1871): II, 630 ("What in tunket are you making such a to-do about it for?") and in *Life* (January 4, 1894): 13/2 ("What in name 'o' Tunket makes all boys so crazy to leave the old farm.?") Matthews, Mitford M., ed. *A Dictionary of Americanisms*. Chicago: University of Chicago Press, 1951.

16. Crowe, W. S. "Small Norway Pines Often Were Lost in Long River Journeys." *Escanaba Daily Press*, Seventh Annual Lake States Logging Congress Souvenir Edition (September 18, 1952).

17. The "new" post office Crowe references was built by the Civilian Conservation Corps (CCC) and dedicated in 1940. The logging mural, which has been re-touched over the years, is still on the wall. (See Page 14.)

18. "Each man wanted to be the best swamper, sawyer, axman, undercutter, or top loader in camp. And each man wanted to work at the best camp, the camp that hauled the most logs . . . jacks bet on their stamina, their bone and muscle. An Iron County jack bet he could fell, limb, cut into four-foot lengths, split and pile a cord of hardwood in an hour. He did it with six minutes to spare . . . every camp made its own 'world's fair load,' a record load of white pine logs on a single sleigh." Bacig and Thompson. *Tall Timber—A Pictorial History of Logging in the Upper Midwest*. 1982, p. 80. The "World's Fair load" refers to a load of giant white pine logs made up in 1893 in Ontonagon County and exhibited at the Chicago World's Fair that year. "A Heritage Built of

Logs." *Milwaukee Journal* (July 9, 1978). (See photograph of the World's Fair log load on Page 36 and on back cover.)

19. The Weston Furnace Company, and the later Charcoal Iron Company at the same site, were located north of Manistique along the river, near the present site of the Road Commission sheds. This location is shown in Orr. *Lumberjacks and Other Stories*, map, p. 4.

20. Horses gradually replaced oxen for hauling loaded sleds out of the woods and on the logging roads. "The problem of shoeing oxen was the main reason . . . the blacksmith could not pick up an ox's foot to put a shoe on it as he could with a horse. If an ox's foot is lifted off the ground, the ox will either lie or fall down . . . elaborate racks had to be constructed which lifted the ox for shoeing, . . . the animal required two shoes per foot, shoes which came off easily when the ox was hauling logs along ice roads." Bacig and Thompson. *Tall Timber—A Pictorial History of Logging in the Upper Midwest.* 1982, p. 30 (with photo).

CHAPTER TWO

21. The village of Steuben, in northern Schoolcraft County, was named for Steuben County, New York, the home of a Chicago Lumbering Company shareholder. Several logging camps were established in the forests surrounding Steuben and it was a center for logging drives down the Indian River to the mills at Manistique. The last river drive to go through Steuben on the Indian River was in 1919. Peterson, Josie. "Logging played an important part in the growth of Steuben." *Manistique Pioneer-Tribune* (March 4,1976).

22. "Today" refers to the late 1940s when Crowe was writing his articles about the lumbering era. White pine was the prized wood for lumber in the years of the Chicago Lumbering Company. According to several sources familiar with 2002 wood values, it is now worth less on the market than when Crowe was writing. One Upper Peninsula lumberman said that if he had a good tract of white pines today, he would leave the trees where they were, due to the lack of profit in cutting but also just to see them growing again.—Jim Schneider, formerly of Schneider Bros. Lumber Company, Marquette; John Hebert, Sawyer-Stoll Lumber Company, Delta County; and Jake Hayrynen, Longyear Company, Marquette, conversations with Emerick, 2002.

23. According to R. A. Brotherton, an early lumberman, 'The bulk of the pine cut in the early sawmills went into wide width, long length boards, inch thick rough sawed timber, joists and 2 x 4s. . . . an important side line of the early days were pickets of white pine for fences in the eighties and nineties." Brotherton, R. A. *Early Logging Days.* Photocopied manuscript. 1950, in collections of J. M. Longyear Research Library—Marquette County Historical Society, and Peter White Public Library, Marquette.

"The white pine was especially prized for building because of its clear straight grain and high durability. The wood was light (floated well), soft and tough, yet weathered with little warping. (The logs) could be made into clear lumber without knots because the limbs started high on the tree." Waring. *Yellow Dog Tales and Logging Trails to Big Bay, Michigan.* 1986, p.6.

24. Ed Ekdahl remembers, "When applying for employment in the forest or the three sawmills, you had to have an 'Employment Certificate' which you signed before you were assigned to a job." Orr, Jack. *Lumberjacks and River Pearls*. Manistique, MI. *Manistique Pioneer-Tribune*, 1979. p. 23.

CHAPTER THREE

25. The Goodrich liner *City of Ludington* brought the author from Chicago to Manistique in 1893.

26. Ed Ekdahl of Manistique recalled that "the vessels that came most frequently were the steamer *Buell* and her two tow barges *Stewart* and *Eleanor* and the steamer *Phals* with one tow barge, the *Delta*. Each boat would take on approximately 200,000 board feet of lumber. During the late summer and fall, it was not unusual to see ten or twelve vessels loading or waiting to be loaded. Shipping would start in April and close the first week in December." Orr. *Lumberjacks and Other Stories*. 1983. p. 73.

27. Lumber schooners called lumber "hookers" were designed for shoal (shallow) water operations. They were wide in the beam and built to carry heavy weight; stone or iron ballast was probably carried at the keel to make them stiffer when sailing empty. A steam tug moved and docked the schooners. Orr. *Lumberjacks and River Pearls*. 1979. p. 13.

Later in the logging period, lumber schooners were replaced by lake steamers and freighters. In 1912, the *Quick Step* of Michigan City, Indiana, the last or one of the very last lumber hookers, was observed sailing out of Ontonagon harbor; the ship sailed on the Great Lakes until 1915. Lumber schooners operating out of western Upper Peninsula ports took loads to Ft. William, Ontario, and also to Tonawanda, New York, and other ports down the Great Lakes. "Passing of 'Lumber Hooker' on Great Lakes Ended Era of Romance in Logging Industry." *Marquette Mining Journal* (February 6, 1954).

28. "The Beavers" was a term commonly used to refer to a group of islands in northern Lake Michigan, including Beaver, Garden, Hog, High and some even smaller islands. Several of these islands are now part of the Beaver Islands Wildlife Management Area—Glen Matthews, Michigan Department of Natural Resources, Gaylord, conversation with Emerick, January 2002.

29. Manistique was a major fishing center in the 1880s-1890s. Tons of whitefish and trout from northern Lake Michigan waters were shipped to Chicago and worldwide, with France being a large market. Fishing tugs still operate from the riverside in Manistique harbor.

30. A description of the construction undertaken by the Chicago Lumbering Company between 1872-80 to facilitate the loading of lumber onto the schooners and later the steamers is found in the *Manistique Centennial Book*. 1960, p. 10. "Tramways were built, one of them between two piles of lumber, with a slip or canal in the middle, deep and wide enough for lake vessels to navigate. Vessels were loaded directly from the piles to save hauling lumber to the docks from storage piles. The firm manufactured 10 to 20 million board feet of lumber monthly, and there were 35 to 50 million feet on the dock at all times."

CHAPTER FOUR

31. "Waterways were indispensable to the early lumbering in the Upper Peninsula. Mills were located at the mouth of the Menominee, Ford,

Escanaba, Rapid, Whitefish, Sturgeon, Manistique, Tahquamenon, Sucker at Grand Marais, Dead, Yellow Dog and Ontonagon Rivers." Brotherton. *Early Logging Days*, 1950. In addition to the last five rivers mentioned by Brotherton, which flow into Lake Superior, Betty Waring notes also "they drove logs down the Laughing Whitefish, Chocolay . . . Salmon Trout, Huron and other rivers, rafting them to mills at Marquette, Pequaming, L'Anse, Baraga and Bay Mills." Waring. *Yellow Dog Tails and Logging Trails to Big Bay, Michigan*. 1986. p. 7.

"Pequaming (in Baraga County) was the first large-scale lumbering and milling operation in the Lake Superior region—boasting over a twenty-year period an annual production average of thirty million feet of lumber. Charles Hebard . . . founded the sawmill and community in 1878. . . . He acquired the site specifically for its deep protected harbor and its access to timber." Clevens. "Henry Ford's Tasty Little Town—Life and Logging in Pequaming." *Michigan History* (Jan-Feb 1999): 19.

Logs were gathered in huge booms and rafted to Pequaming and Hancock from as far away as Grand Marais—a float of 150 miles along the south shore of Lake Superior. Hesterberg and Noblet. "Annual Spring Drives on Copper Country's Mighty Sturgeon River Recalled." *Marquette Mining Journal* (June 15, 1957).

32. According to Harvey Saunders, pioneer lumberman, the Alger, Smith Company, operating as the Manistique Lumbering Company, ran logs from its own cuttings on the Fox River and other northern sites to the mills at Manistique during the earliest years of logging despite the long river distance; the cost of shipping logs on Lake Michigan was $2.00 per M board feet lower than on Lake Superior. However, when the mill scale of Alger, Smith logs was consistently lower than had been figured in the woods, a survey of the log ponds was done and Alger, Smith logs were found in the ponds of several other companies. The company extended its Manistique Railroad to Grand Marais; thereafter most of Alger, Smith's large cut was sent by rail to Grand Marais. The logs were rafted down Lake Superior or transported by schooner. Saunders, Harvey. "Timber, Trains and Alger-Smith." *Great Lakes Pilot and Pictured Rocks Review* (September 8 and 22, 1971); Carter, James L. "Burt Township." In *Alger County—A Centennial History 1885-1985*. 1986.

33. Of the dams referenced by Crowe: Hartney Dam was built where the Indian River enters Doe Lake; Harrigan Dam was a half-mile south of Hovey Lake; Little Tote Road Dam was near the present Mirror Lake: Six-Mile Dam was located at the present Widewaters on the Indian River and flooded Fish Lake and upstream; Ten-Mile Dam was located where Forest Road 2258 crosses the Indian River and flooded the present Chain of Lakes—Deep, Corner, Ostrander, Straits and Skeels Lakes. Six- and Ten-Mile dams were both built in 1888 and raised the water about ten feet. Jackpine Dam was located near the site of the present Jackpine Bar on M-94. A dam was also built at Steuben which backed the waters of the Indian River over thirty acres. In 1900, a spring flood washed out all the dams except the one at Steuben. Ten-Mile dam was rebuilt, and a new dam with a bridge was constructed where the river leaves Doe Lake, but logging was over above Doe Lake and those dams were not rebuilt. *Alger County—A Centennial History 1885-1985*. 1986.

34. Fred Rydholm, U.P. historian, notes that when loggers first came to the area in the mid-1880s, their saws had no "raker teeth" (to rake the sawdust from the cut). "Without raker teeth, the saw would bind up while slicing through white pine, slowing work to a crawl. . . . Blacksmiths discovered that by removing every other tooth, the sawdust would be raked out of the cut. Finally raker teeth were invented, along with the crosscut saw. Then two men could saw straight through the big trees without a stop." Goodrich, Marcia. "Northwoods Progress." *Marquette Mining Journal* (September 14, 1991).

35. The landlookers, even while deep in cedar swamps and beaver meadows along the rivers, could at times identify white pine tracts by the sound of the wind in the tops. William Davenport Hulbert gives a vivid word-picture of "the landlooker and the naturalist" who hear the white pines trees long before they break out into the grove. Beeson, Lewis, ed. *White Pine Days on the Tahquamenon.* Lansing: Historical Society of Michigan, 1949.

36. The largest tree as described by Crowe was located in Section 14, T47N, R16W in Hiawatha Township, northern Schoolcraft County. The "Wolf Lake" referenced by Crowe is shown on 1930s-era maps; the lake has also been called Round Lake. It is now named Cusino Lake, earlier spelled as Cousineau. Another small lake farther north in Alger County is now designated as Wolf Lake. The largest tree is also noted in *Michigan History* as measuring 7 1/2 feet in diameter at the stump, with five prongs which cut into 16 logs, 16 feet long. Leech, Carl Addison. "Pictures of Michigan Lumbering." *Michigan History* (Autumn 1939): 339.

37. Jim Schneider recalls logs cut in Marquette County which would not fit through the six-foot by six-foot-square doors of the sawmill and had to be split lengthwise before sawing—Schneider, conversation with Emerick, February 2002.

38. The Tahquamenon River, Luce County, in the eastern Upper Peninsula, is remembered as one of the harder rivers to drive. The 48-foot-high upper falls was a formidable barrier to floating the logs down to Lake Superior. . . . "old growth timber flowed like pick-up sticks over Tahquamenon Falls. Floating trees broke like matchsticks when they crashed onto the splintered timbers piled at the bottom of the Falls." "Log Mining Proposal Stirs Debate." *Marquette Mining Journal* (October 1, 2000).

Some drives were successful. Loggers persisted, and in 1893 Dan McLeod "took out square timber for the Flatt Brothers of Hamilton, Ontario, near the Tahquamenon River. The sixty-foot timbers were floated over the falls and down to the mouth of the river where they were loaded into boats to be carried to England." The story is that much of this timber went into the construction of Buckingham Palace. Gerred, Janice. "When White Pine was King." *Grand Marais Pilot and Pictured Rocks Review* (June 2, 1989).

Con Culhane was a well-known logger in the Tahquamenon country. He met the challenge of driving logs over the Tahquameon Falls by driving during low water in August. His men tied ropes around their waists and beveled the rocks at the lip of the falls with chisels and sledges so the logs would shoot over the edge instead of dropping straight down to become splintered or tangled. Culhane's logs—an estimated 100 million board feet—were then driven to the Chesborough mill at Emerson. McTiver, Barnard. "Con Culhane, Real Life Paul Bunyan." *Michigan Conservation Magazine* (Jan-Feb 1954): 11-14.

William Davenport Hulbert, in *White Pine Days on the Tahquamenon* (1949), provides vivid portrayals and photos of logging along the Tahquamenon River. The Logging Museum at Newberry displays photos, tools and artifacts of logging days, along with a reproduction of a Civilian Conservation Corps camp, a cook camp and a nature walk along the Tahquamenon River.

39. One who worked in the lumber camps remembers . . . "(while) I was in the forest, employed as a clerk and scaler, very few of the lumberjacks ever asked what the pay was; they were more concerned about the food so they always asked, 'Who's the cook?' The wages at that time in the forest such as log cutters, skidding, log decking, etc., were thirty dollars a month. The cooks received fifty and the teamsters (with their horses) forty-five and that included their board." Orr, quoting Ed Ekdahl. *Lumberjacks and River Pearls*. 1979, p. 24.

40. The $15/day of the late 1940s, cited by Crowe as equivalent to $2.50/day at the turn of the century, would equate to $193/day in 2002. However, lumberjacks usually were employed in the woods for only part of each year, from late fall to late spring. If a man worked solely as a river driver, his seasonal employment was even more brief. Some lumberjacks found work in the sawmills during the summer months. Monetary calculations from the website: *http://www.minneapolisfed.org/economy/calc/cpihome.html*

41. The Stutts Creek drive began about three miles east of the present M-94 and ran to the mouth of the Manistique River. The Stutts was considered a difficult stream to drive because it ran through a flat swamp; when logs jammed up or drifted back into the swamp, the natural water flow would be lost. *Alger County—A Centennial History 1885-1985*. 1986.

42. Logging and drive camps were identified by number and also by name of the foreman. Many camp numbers have been interchanged and transposed over the succeeding years. Harvey Saunders, in a letter to Crowe dated January 20, 1952, notes that he is forwarding a map on which he has marked Schoolcraft County camp locations. That map survives and with other documentation, gives evidence of a few of the C. L. and W. L. camps. Camp 60 was built in 1897 at Round Lake (also named Wolf Lake, Cousineau and later Cusino Lake) and operated for three years until the cut of white pine was gone; Camp 47 of George Hovey was at Hovey Lake; Camps 35 and 65 were near Steuben; Camp 70 was at Doe Lake; Camp 76, another Hovey camp, for the upper Indian River log drive, was located near Ten-Mile Dam on the Chain of Lakes; Camp 79 at Cookson Lake was managed 1903-1913 by Frank Cookson; Julius Phillion managed Camp 81 at Klondike Lake and Camp 85 was ten miles northwest of Steuben where the last tract of large white pine of the Chicago Lumbering Company holdings was cut by Consolidated Lumber in the winter of 1914-15. The C. L. and W. L. companies typically ran nine camps per season. *Schoolcraft County map*. Manistique, MI: Schoolcraft County Road Commission, 1954; Carter, James L. *The Cusino Story, A Brief History*. Marquette, MI: Northern

Michigan University, 1972; Wm. S. Crowe correspondence.

43. In 1906, Harvey Saunders set up a lumber camp five miles below Germfask on the Manistique River where he "peeled hemlock bark in the summer and logged in the winter." He also had a camp during 1915-1917 near the Fox River in T46N, R13W, Section 14 northeast of Germfask and a photo of that camp survives. (See Page 51.) Saunders continued as foreman for the Chicago Lumbering Company and its successor, the Consolidated Lumber Company, until 1919 when the cut of timber was finished. Saunders. "Timber, Trains and Alger-Smith." *Grand Marais Pilot and Pictured Rocks Review* (August 26, 1971).

44. Harvey Saunders was interviewed by William S. Crowe on several occasions for these reminiscences. According to Saunders, "I read in the *Pioneer-Tribune* that Will Crowe's *Lumberjack* book is going to be printed soon. I think it will be good as I talked into a dictaphone for him for 6 hours, and many short conversations to get material for his book." Saunders, Harvey Cookson. *Jamestown* (undated manuscript), Wm. S. Crowe collection.

45. For fourteen months, beginning in April 1944, Saunders also was foreman of a World War II Conscientious Objector's Camp, located at the site of an old CCC camp south of Germfask, in eastern Schoolcraft County. "Saunders is Given Honor," *Manistique Pioneer-Tribune* (April 4, 1957).

CHAPTER FIVE

46. Decoration Day is the present-day Memorial Day

47. There were actually more newspapers during some of the years Crowe describes. The first newspaper in the city was the *Manistique News*, started by Thomas MacMurray, grandfather of screen actor Fred MacMurray. Among the county-wide papers, Major W. E. Clark's *Schoolcraft Pioneer*, started in 1880, was at various times a semi-weekly and tri-weekly newspaper; it merged with the *Manistique Tribune* in 1896, after Clark's death. The *Record* was started in 1904 by Ben Gero. The *Härold*, published by Nettie Steffenson, and the *Courier* of John McNaughton were also publishing by 1904. By 1922, the merged *Pioneer-Tribune* was the only weekly paper still operating. *Manistique Centennial Book*. 1960.

48. The *Harper's Weekly* article referenced by Crowe poked fun at the wandering cows of Manistique, which, though fairly well tolerated by the populace in the past, had taken to pilfering quantities of vegetables from local markets, resulting in passage of the "cow ordinance" by the city fathers. An excerpt: "It was decided that cows must keep off the streets at all hours, and it is now possible to buy fresh vegetables in Manistique, and ambitious young heifers no longer put their heads in at the general delivery of the Post Office, to the consternation of the young lady in charge. The cows of Manistique abused their privileges and lost them. Many very able politicians may leam a valuable lesson from this simple experience of the Manistique cows." Carruth, H. "Afield with Fact: Downfall of Cows in Manistique." *Harper's Weekly* 40 (December 12, 1896): 1219.

49. Jenny Putnam first came to Manistique in 1888 and returned in 1890; she described her first sight of Manistique: "To one from 'outside' or 'down below', the town looked queer. The streets were covered with big stones, there were no trees, scarcely any grass or flowers, cows were running loose, there were great piles of lumber and over all the agreeable odor of pine. Like many other towns of that time, the sidewalks were of boards laid lengthwise . . . Nothing was safe from the cows and chickens. When the new housewife put her box of parsley on top of the woodpile for safekeeping, it was promptly devoured by hungry bovines...the cows also helped themselves to the grocer's display of greens." Orr. *Manistique Pioneer-Tribune* special edition. In *Lumberjacks and Other Stories*. 1983. p. 7.

50. "Horse Heaven" is about ten miles south of Prosser, WA. *Rand, McNally and Company Indexed Atlas of the World.* Chicago: Rand, McNally and Company, 1892. p. 359.

51. Will Crowe was an active participant in bicycle races throughout the area. In a 1937 article for the *Manistique Pioneer-Tribune*, he remembered the bicycle times: "The only means of travel for long distances was by railroad train or boat and horse-and-buggy for trips in the country or to neighboring towns, but the bulk of local travel was on foot, and the vast majority of the people wouldn't get more than 10 miles from town in a year and then—along came the bicycle . . . is it any wonder that everyone went bicycle crazy? In the evenings, bicycle processions would form and people would line the sidewalks to watch the 'scorchers' and impromptu races. Bicycle races were held on Decoration Day, 4th of July, Labor Day and at specially-promoted race meets.

"Century clubs were formed by riders who had ridden 100 miles in a day . . . . The writer purchased a 14 pound *Thistle* in 1894 for $110.00, rode it for two years and sold it . . . for $80.00. In 1895, I rode this bicycle from Green Bay to Milwaukee via Oshkosh, a distance of 145 miles over country roads in 14 hours." "W. S. Crowe, Bicyclist of 90's Recalls Days of Local Club." *Manistique Pioneer-Tribune* (October 28, 1937).

52. Potatoes not only grew well—and still do—in the Upper Peninsula, but were celebrated in their own show. In a "Memories" column in the *Manistique Pioneer-Tribune*, August 26, 1982, Jack Orr includes a 1940 photo of the U. P. Potato show and adds "Potatoes were a big thing in the history of our county and local producers were renowned in this specialty." In 1937, Alphonse Vershure of Hiawatha Township, Schoolcraft County, with a yield of 523 bushels per acre was the sweepstakes winner at the U. P. Potato Show, held at L' Anse in Baraga County. The 1940s marked the peak of potato production in Schoolcraft County and in the Upper Peninsula with almost 25,000 acres in production. By 2001, there were two dozen commercial potato farms in eight Upper Peninsula counties; only about 3,500 acres are planted to potatoes, but the "farm value" is about seven million dollars—Ben Kudwa, Michigan Potato Industry Commission, and Chad Cloos, Michigan Agricultural Statistics Service, conversations with Emerick, February 2002; "Michigan Potato Facts." *Marquette Mining Journal* (January 27, 2002.)

53. Manistique High School was opened in 1892; the first class of Oren Quick and George Pembroke Tucker graduated in 1894—Margaret Cain, Manistique resident and former Manistique schoolteacher, and Vonciel LeDuc, president, Schoolcraft County Historical Society, conversations with Emerick, February 2002.

54. "At the turn of the century, (Crowe) was one of the top pitchers in indoor baseball in the country. At what is now the Oak Theater, he pitched the game in which the Olympic Club of Manistique defeated the crack Spaulding team of Chicago, a team made up largely of professional players." "Former Banker, Civic Leader, Athlete, William S. Crowe Succumbs Here at 90." *Manistique Pioneer-Tribune* (November 4, 1965).

55. Kitch-iti-ki-pi is the Indian name for the Big Spring. In legend, it is named for a young Chippewa chief from a lodge on the east shore of Indian Lake. Now officially Palms-Book State Park, the Big Spring is a clear-water natural spring, sixty feet deep, four hundred feet across, with an outflow into Indian Lake. The spring has a constant water flow of about 10,000 gallons per minute at a year-round temperature of 45°. In her recollections of Manistique in the 1880s, Jennie Putnam described the way to reach the Big Spring in the days before a railroad. "The Big Spring at that time was reached by a long row (by boat) across Indian Lake and then poling up the lovely little creek. This was a fitting approach to such wonderful beauty." Orr. *Lumberjacks and Other Stories*. 1983. p. 8; *A Souvenir of Manistique, Michigan. Manistique Härold*, 1902; reprinted by Schoolcraft County Historical Society, 1992.

56. The couplet is from the poem "The Mary Gloster," by Rudyard Kipling, 1894.

CHAPTER SIX

57. Wilfred Nevue was a native of Republic (Marquette County), Michigan, and worked in timber camps in the Upper Peninsula, Minnesota and Wisconsin. In 1954, his article for a local newspaper was headlined "Incredible Lumberjack Stories Debunked by Veteran of Camps." Nevue tells of an encounter in a railway station with a man posing as an old-time woodsman and telling stories of terrible lumberjacks and immoral towns. The storyteller gave lurid details of Nevue's own home town, describing buildings and people which did not exist. Nevue concludes: "From such blow-offs has come the stuff from which salacious tales about lumberjacks have been gotten and are being fabricated." And "In 1935 a nationally known publisher, the Arthur H. Clark Company, stated in an advertisement card concerning books on lumbering, 'recently several volumes have been published, written from hearsay, full of exaggerations and incorrect from a geographical and biological standpoint.'" *Marquette Mining Journal* (February 13, 1954).

Following the publication of his book, Crowe received a letter from Courtney E. Carlson of Newberry, Michigan. Carlson writes, "I see you have read the book 'Call it North Country' and apparently consider it highly colored. I share your views. You will note that, in the chapter about Seney, it makes passing reference to a 'man named Carlson' who ran a boardinghouse there. That man was my grandfather . . . my father was a logger in the pine days and used to tell me about it." Carlson, Courtney E. Letter to Wm. S. Crowe, May 23 (no year). Newberry, MI.

58. Crowe refers to John Bartlow Martin, author of *Call it North Country*, first published by Alfred Knopf, New York City, in 1944.

59. Ed Ekdahl, remembering those early lumbering days, concurs: "When they came into town and got a little drunk and happened to meet a woman on the street, they would step off the walk, take off their hat and say, 'Pardon me, lady,' and they never used obscene language or made offensive remarks in the presence of women. My eighth grade teacher, who came from Boston, commented on the behavior of the lumberjacks and I quote, 'They are very courteous men, far more so than our Bostonians.'"

Orr, quoting Ed Ekdahl. *Lumberjacks and River Pearls.* 1979, p. 24.

60. Dollarville, the site of a sawmill established by Robert Dollar on the Tahquamenon River, is in Luce County in the eastern Upper Peninsula. Dollar found fine stands of white pine in northern Schoolcraft County and conducted logging operations on the Autrain and Laughing Whitefish Rivers, and farther east on the Two-Hearted and Tahquamenon Rivers. Karamanski, Theodore J. *Deep Woods Frontier, A History of Logging in Northern Michigan.* Detroit: Wayne State University Press, 1989. Original source: Dollar, Robert. Diary (typewritten manuscript). Bancroft Library, University of California, July 9, 1882, and September 27, 1886.

61. Michigan's original forest covered 35 million of the state's 38 million acres . . . more than a quarter of the original Michigan forest was made up of . . . white pine. Worth, Jean, in Panax Newspaper Corporation Newspaper Group (*Escanaba Daily Press*) (July 1, 1976).

CHAPTER SEVEN

62. "By the end of the lumbering era, Michigan loggers will have cut 161 billion board feet of pine logs and 50 billion board feet of hardwoods. . . . In dollar value, Michigan lumber will outvalue all the gold extracted from California by a billion dollars." "Michigan Through the Years: A Brief History of the Great Lake State." Lansing, MI: Michigan Historical Center, Michigan Department of State, n.d.

63. Wheeler, W. Reginald. *Pine Knots and Bark Peelers; the story of five generations of American Lumbermen.* La Jolla, CA: Privately published, 1960.

64. Corinne, Michigan, was a lumber settlement in Newton Township of Mackinac County, in the eastern Upper Peninsula. It was given a station on the Minneapolis, St. Paul and Sault Ste. Marie Railroad but to avoid confusion with Corunna, an established town in the Lower Peninsula, it was given a post office named Viola in 1889; the office operated until 1942. Romig, Walter. *Michigan Place Names.* Detroit: Wayne State University Press, 1986.

65. The Little Wind River is in Southwest Washington in Skamania County; it is a tributary of the Wind River, which flows into the Columbia River. An area of mountains and forests, this would be a logical place for an early timber claim. The Timber and Stone Act of 1878 (Forty-Fifth Congress, Sess. II, CH. 151) allowed individuals to place claims on 160-acre timber tracts in the Northwest states for $2.50 per acre. Subject to valid existing rights and claims, the act was repealed in 1955. Carol Green, Head, Forest Resource Center, University of Washington, Seattle, electronic communications to Emerick, January, 2002; *United States Statutes at Large* (Public Law 206, Chapter 448, HR4894, August 1, 1955). Washington, DC: U.S. Government Printing Office, 1956.

66. Crowe wrote that the cruises of both John Lyberg, the "expert" for the Union Trust Company, and the cruises contracted for the Consolidated Lumber Company in the summer of 1912, "showed sixteen million feet of pine, including eight million feet of virgin white pine . . . I still have copies of the cruises if anyone wants to see them. I also have the Chicago Lumbering Company's old 'Log Book' showing the number of logs and log scale of all logs put in, jobbed, or purchased by the Chicago and Weston Lumber Companies from 1884-1885 to 1900-1901, inclusive, beginning with Camp 7 and ending with Camp 73."

Crowe, Wm. S. *Lumberjack*. Manistique, MI: Privately published, 1952. p. 36.

67. The Chicago Lumbering Company owned most of the homes and businesses in Manistique and rented them to employees and others for $4.00 to $8.00 per month. When an employee left the company, his family would no longer be able to rent the house and housing was difficult to find for non-employees. When Crowe and Yalomstein founded the Consolidated Lumber Company, "the new company made homes and businesses available for private purchase by individuals." *Manistique Centennial Book*. 1960, p. 12.

68. The reference is to W. J. Murphy, publisher and owner of the *Minneapolis Tribune* newspaper, who organized the Manistique Pulp and Paper Mill on the river in 1916. The mill opened in 1920 after several delays. The water purchased from the Consolidated Lumber Company was used in the manufacture of paper at the mill. *Manistique Centennial Book*. 1960, p. 13.

69. The location where Consolidated cut the last of the big pines of the Chicago Lumbering Company is on the border of what is now the Big Island Wilderness Area within the Hiawatha National Forest.

CHAPTER EIGHT

70. "To date" indicates the late 1940s to early 1950s—the time that Crowe was writing and publishing his articles and book.

71. The reference is to the canal and locks at Sault Ste. Marie, built to allow ship passage from Lake Superior to Lake Huron through the rapids of the St. Mary's River. During 1845-1855, the business of portaging freight and ships around the rapids gave employment to most of the men in Sault Ste. Marie. Several large locks were built during the time Crowe describes, including the MacArthur Lock, opened in 1943.

Manse, Thomas. *The Soo Locks*. Sault Ste. Marie, MI: Privately published, 1967.

72. There were several specialized manufacturing operations in the peninsula, using pine and hardwoods as raw material. The Lake Independence Lumber Company, in Big Bay, Marquette County, later purchased by Brunswick in 1908, made blank bowling pins, clothespins and lumber for General Motors. For the four different kinds of bowling pins made there, rough blocks of green hard maple were air dried, kiln dried and made into finished pins. Waring. *Yellow Dog Tales and Logging Trails to Big Bay, Michigan*. 1986, pp. 19-21.

The Manistique Handle Company was another large user of hard maple; built in 1912, the company manufactured handles for brooms and feather-dusters along with dowels. Culls were made into toy floor-sweeper handles and whisk brooms. Later beech logs were used for heavy warehouse broom handles. Monthly production was 35,000-40,000 handles; most were shipped to Chicago and St. Louis broom companies. Some of the handles were also shipped to various prisons and used in the manufacture of brooms. Orr, quoting Ed Ekdahl. *Lumberjacks and River Pearls*. 1979, p. 34.

"Many thousand feet of clear white pine went into the manufacture of matches . . . in the Ontonagon district. Diamond Match Company operated extensively for many years." Brotherton, R. A. "Early Logging Days." *Ishpeming Iron Ore* (v. LXXI, No. 18, December 24, 1949). (The company) "made 2 1/2 inch pine blocks, which were forwarded to Oshkosh, WI, to manufacture into wooden matches." "Almost 80 years since Ewen's Big Pine Load was in Chicago." *Houghton Daily Mining Gazette* (December 9, 1972).

73. " . . . cedar shingles, lath, posts, poles, ties and cedar lumber for boat construction were in great demand. Cedar blocks 6" thick were manufactured for street paving . . . streets in Menominee, Marinette, Escanaba and Gladstone boasted of this fine new pavement."

Brotherton, R. A. *Ishpeming Iron Ore* (v. LXXI, No. 18. December 24, 1949).

74. An early land report agrees: "A very large portion of the Northwestern States (referring to Illinois, Iowa and other territories) is prairie country, without timber or lumber for the commonest purposes. This region . . . is growing in population, wealth and general prosperity faster than all the rest of the Union together, and is to a very great extent dependent on Michigan for its lumber." *Pine Lands of the Saint Mary's Falls Ship Canal Company*. Detroit Land Office, 1857, p. 11.

As the railroads pushed west through the prairie states and to the West Coast, great quantities of ties and trestles were needed and were supplied with wood from the northern logging camps and Great Lakes mills. Waring. *Yellow Dog Tales and Logging Trails to Big Bay, Michigan*. 1986, p. 5.

75. Brown, Andrew H. "Work-hard, Play-hard Michigan." *National Geographic* (March 1952): 279-320.

76. The Upper Peninsula of Michigan "is to be credited with 1,154,844,355 feet of lumber and 550,833,000 shingles or one-eighth of the total white pine production of the entire northwest and of the shingle production." "A Magnificent Showing—Product of the Saw and Shingle Mills of the Upper Peninsula for 1892." *Marquette Mining Journal* (February 10, 1893). In Michigan Lumbering Collection of the Peter White Public Library, Marquette, MI. In that year, the Chicago Lumbering Company of Manistique is credited with cutting 85,000,000 board feet of lumber, with 16,000,000 on hand. The production of eighty-eight different Upper Peninsula saw and shingle mills operating in 1892 is detailed in this article.

Glossary

BAGNIO—Bordello; house of prostitution.

BAND SAW—A saw in the form of an endless belt running over pulleys.

BEAT—During a river drive, the river was divided into sections, and a small crew would handle each section, or beat, of the drive. As described in *Lumberjack*, such a section would be called by the name of the head driver, as in "Moran's beat."

BIRL—To rotate a floating log by treading upon it with calked boots; done to find the water mark by working rivermen, and performed as a stunt or contest for entertainment by modern birlers.

BLOW IN, also **BLEW IN,** also **BLOW HER IN**—Spend all of your stake (pay) in town, usually on one big spree after a winter's work in the woods.

BOLE—The base end of a tree.

BOOM—Logs chained together at the ends to form a corral to hold logs in the water together until ready for refloating. reshipping or sawing. A group of logs pulled together by heavy boom sticks for transportation by water.

BREAK-UP—Closing of a logging camp at the end of a season in spring of the year. Also ice melting on roads and rivers due to warm spring weather.

CALKED BOOTS, also **CORKED BOOTS**—Boots with short, sharp spikes set in the soles. River drivers used these boots to keep from slipping off the logs when they had to ride them.

CAMP ORDER—The only difference between a camp order and a time check was that a camp order was issued for a certain flat amount, say $5.00, $10.00 or $15.00, against a lumberjack's pay. The purpose of camp orders was in case a man wanted to buy something while at the camp. Jewelers and clothing merchants visited the logging camps and took camp orders in payment for merchandise.

CANT—Hewn timbers or a timber squared in a sawmill.

CANT HOOK—A tool like a peavey, but having a toe ring and lip at the end instead of a pike. Used to handle logs on land, or on river drives for dislodging logs from the banks and in breaking jams. Once described by a greenhorn as a stick with a hook hanging on the end. A classic remark heard on a drive when a riverman fell in the river was "To hell with the man, save the cant hook."

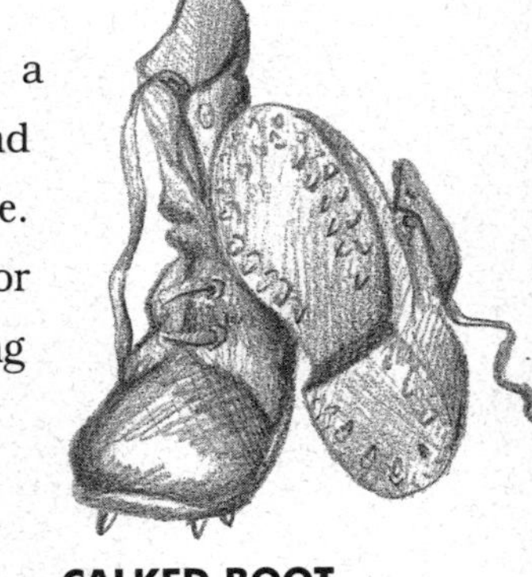

CALKED BOOT

CAYUSE—A native range horse.

CEDAR SAVAGE—One who cuts or peels cedar logs, poles or posts.

CHAFF—Something considered worthless; comes from the seed coverings and other debris separated from the seed in threshing grain.

CANT HOOK

CHAUTAUQUA—Adult education courses featuring lectures, music and dramatic presentations, performed by traveling artists. The Chautauqua movement is named for the western New York county and lake where it began as a religious training school for adult educators shortly after the Civil War. The related Lyceum courses began in 1826 in Massachusetts.

CIRCULAR SAW—A rotating steel disc with cutting teeth around the edge.

COOKEE—Any kind of cook's helper in the logging camps.

COOPERAGE—Plant or factory to make wooden barrels, casks or tubs. A person who makes such wooden ware is a **COOPER**.

CORD—A stack of wood four feet high, four feet wide and eight feet long. It equals 128 cubic feet and contains about ninety cubic feet of solid wood and bark.

COUPONS—Coupon books were issued at the main office of the company at any time to any employee who had the credit coming. These coupon books were issued in denominations of $1.00, $2.00, $5.00, $10.00 and $25.00, and inside the book were printed coupons ranging in value from 1¢ up to $1.00 just like postage stamps. The coupons were good at any of the company's stores and at other stores in town. Any coupons taken in by other merchants were redeemed in cash by the company on a monthly basis.

COURSE—A single stack of boards, usually all the same size. Several courses piled adjacent and bound by crossers would make a pile with a certain size front.

CRIB—Heavy log structure built on the ice and filled with stones, then sunk to the river bottom with tip projecting to form anchors for booms for sorting or holding logs.

CROSSCUT SAW—Saw used in felling and bucking trees. Before rakers were put in saws, only axes were used to cut down timber.

CROSSER—A piece of lumber laid crossways between courses.

CUTTER—A small sleigh.

DEADHEAD—A water-soaked log lying on the bottom of a river or lake, or a partly sunken log.

DINKY LOCOMOTIVE—A small logging locomotive.

DOYLE RULE—One of the common log rules in use in the pine era of the late nineteenth century.

DUMP—A place on a riverbank where logs were dumped from logging trains or sleds into the river or mill pond.

FEATHERBEDDING—The practice of requiring the employment of more workers than are needed for the job.

FLATIRON BLOCK—The block of land bounded by Arbutus Avenue (then Water Street) to the southwest, Pearl Street to the east, Main Street to the north and the old Chicago Lumbering Company yards to the west. Originally owned by Alex Richards, it became the site of several saloons under the development of Dan Heffron, who became an associate of notorious Seney saloon keeper Dan Dunn. The Flatiron Block was one of the few parcels of land in Manistique not owned by the C. L. Co., whose directors strongly opposed the development of the saloons and club-rooms. Although his brother Dennis was once county sheriff, Dan Heffron was eventually arrested and, when found guilty, made an escape from the city by horse-drawn sleigh.

FORTY—Smallest unit of acres in which timber is traded. A subdivision used in timber survey consisting of one-sixteenth of a section. Forty acres of land or timber.

FRONT—The end of a stack of lumber in the mill yard, facing the loading dock.

GANG SAW—A series of saws arranged parallel in one saw rig for sawing lumber.

GAR—Grand Army of the Republic, the Union Army in the American Civil War.

GERMAN SOX—Warm woolen lumberjack socks.

GIBSON GIRL—An idealization of the American girl of the 1890s as portrayed by Charles Dana Gibson.

HAYWIRE—Bankrupt; "belly-up." A nickname for the Manistique & Lake Superior Railroad. A term applied to any outfit that operated "on a shoestring" and was considered to be "held together with baling wire."

HIGH ROLLWAYS or **ROLLAWAYS**—Banks of the Manistique River where the logs were stacked and then rolled into the river in the spring. The place where the riverbanks were the highest.

JACK LADDER—Heavy endless chain with dogs (metal studs) which carry logs out of the mill pond up to the gate saws.

JOBBER—A small logging contractor.

KERF—The slit or opening made by a saw; the width of the saw cut.

KLONDIKE—A boarding house on the old State Road near its junction with present M-94; it served as Manistique's only brothel.

LANDLOOKER—A man who estimates standing timber. Classic portrayals are found in W. D. Hulbert's *White Pine Days on the Tahquamenon*. and in John Munro Longyear's *Landlooker*.

LIME KILNS—Large stone ovens in which limestone was burned to purify it.

LOG JAMMER—A river driver who watched for and helped break up log jams.

LOG MARK—A symbol stamped in the end of a log with a stamping hammer or chopped into the side of a log with an axe to indicate its ownership. Same as **LOG BRAND**. Used when logs of more than one firm were sent down the river on the same log drive.

LOG SCALE—The board feet content of a log or of a number of logs considered collectively; logs were usually scaled in the woods and sometimes the scale number was marked on the end of the logs.

LYCEUM COURSES—A series of artistic presentations by traveling companies of musicians, dramatists and other artists; the winter equivalent of the Chautauqua. Named for the place where Aristotle lectured his students.

M—A thousand board feet of lumber, logs or standing timber.

MAIN RIVER—What the Manistique River was always called by lumbermen in the Schoolcraft County area.

STAMPING HAMMER

PEAVEY

MANISTIQUE—In early days, the town was known as Epsport, honoring the family of Mrs. Charles T. Harvey. The current name, an Indian word as written by a French Canadian, was originally spelled Monistique. Various meanings have been assigned to the word: "much sand in that place" or "River with the Big Bay" or "Vermillion." Large sand dunes left by glacial action are found here. The name was given to the town on the north shore of Lake Michigan and also to a major river used for logging operations. When the town name was registered with the state, it was misspelled as "Manistique" and that spelling has been retained.

MILL SCALE—The scale (board feet content) of logs delivered to the mill.

NORDEN—"A sick benefit society . . . organized among the Scandinavians in Manistique, Michigan, May 12, 1895, for the purpose of helping one another in case of Sickness and Death," according to a document retrieved from the cornerstone of the Schoolcraft County Courthouse after it burned in 1974. The 1901 document was signed by 116 members. The society numbered 48 in 1895. Sick benefits paid by the society were $5 per week; it paid a $50 burial benefit. "The yearly Dues each Member has to pay is $4.00 in quarterly payments of $1.00 and 50 cents, to help pay Burial Expenses, when the number of Members exceeds 102, and when less $1.00."

NO-SEE-UM—A biting midge (insect), reportedly from American Indian term for "you don't see them."

PEAVEY—Similar to a cant hook, but with the end armed with a strong, sharp spike. For rolling and handling logs in water. Named after Joseph Peavey of Stillwater, Maine, who invented it in 1858. His gravestone in Bangor, Maine, includes a carving of two crossed peavies.

PEERLESS—Tobacco. A staple for the men in a lumber camp. It was smoked in pipes and chewed; sometimes the pipe ashes were used as snuff after smoking.

PENSTOCK—A conduit or pipe for conducting water; also, a sluice or gate for regulating a flow (as of water).

PIKE POLE—A long pole, twelve to twenty feet long, with a sharp spiral spike and hook on one end, used to handle floating logs.

PLANING MILL—A mill for smooth surfacing rough lumber.

PLANK—To prepare fish by baking it on an untreated hardwood board, which was also used to serve the fish.

PLAT BOOK—A detailed map showing ownership and acreage of every parcel of land within the county.

POLE AX—A single-bit ax for chopping and driving wedges and for releasing a crosscut saw from binding when trees were being felled. Used in felling trees in early logging.

POLE TRAIL—Walkway on a log or stack of logs or logjam in a river. Also, a road like a railroad track built of ten-inch or larger logs, doweled together at the end and roughly dressed on top to fit iron concave wheels on which logs were carried to the river or the landing on railroad-type cars. These roads were generally built over swamp areas. The logs were greased, particularly at the corners, for ease of handling.

PULL-UP—A place on the bank of a river, lake or pond where logs that had been cut and floated down out of the woods were pulled up out of the water to be milled into lumber or loaded on a train for shipment to a mill.

RED HORSE—Salt beef; corned or pickled beef.

RIVER DRIVE—Taking logs to a mill by floating them down a river.

RIVER DRIVER, also **RIVER HOG**—One who worked on a log drive.

ROLLED—A lumberjack robbed of his pay or "stake."

SAWYER—One who controlled the carriage and other machinery in sawing logs into lumber. The quantity and quality of the lumber depended on his judgment.

SCALE—To ascertain the number of board feet in a log. To keep a tally of total board feet of measure in logs.

SCHOOLCRAFT—The county was named for Henry R. Schoolcraft, author and Indian agent, who was in Michigan 1820-1842. Originally joined with Marquette County, it was separated in 1871 and included part of what is now Alger County.

SCOOT—An inferior and practically worthless piece of hardwood lumber.

SCRIBNER RULE—A log measuring rule invented by J. M. Scribner in 1846; today known as the Scribner decimal C.

SHINGLE BOLT—A short cedar log, usually eight feet, to be used for making shingles.

SHOE PACS—Footwear with rubber bottoms and leather tops. These were worn in cold weather with several pair of socks.

SHORT STUFF—All bolt-length (100 inches) timber, including shingle bolts, tie cuts and pulpwood sticks.

SLASHBOARD—Dropleaf in a dam, used to release water for a log drive or to build up water behind the dam.

SLASHINGS—The limbs, tops and unused logs left after a tract of land had been logged. It was like tinder when forest fires started.

SLIPS—Artificial piers serving as a landing and loading place. In Manistique, the slips are a specific feature of the Manistique River just downstream from the present paper mill and upstream from the yacht basin. The river's present seven slips (two extending from the west bank, two from the east, and three islands) were completed by 1925, and shown on a U. S. Army Corps of Engineers map of the harbor. An 1873 map by the Corps of Engineers shows only two slab docks on the east bank of the river, in the same area. An 1880 Corps map shows three slips, all in the same area, which was then adjacent to the Chicago Lumbering Company's mill yards.

SORTING GAP, also **SORTING JACK**—A raft or rafts, secured in a stream, with an opening through which logs passed to be sorted by their marks and diverted into pocket booms of the downstream channel.

PIKE POLE

SPLASH DAMS—Small dams built to provide a "splash" of water when needed to float logs. Built of materials at hand—logs, rocks, gravel—with shovels and picks by men, using horses for hauling when available.

STAMPING HAMMER, or **MARKING HAMMER**. Log marks were required to be stamped into the end of each log. The cast iron stamping hammer was crafted with a raised mark on one end; by swinging the hammer like an ax, lumberjacks stamped the mark three or four times in different locations on the log end so that the marks were visible when the log was afloat.

TURKEY

STAKE or **STAKES**—The money earned by a lumberjack while working in the woods; usually paid to him at the end of the season when he was returning to town.

STUMPAGE—The value of timber as it stood uncut in the woods; the value of the timber without the land. The price paid by loggers to a land owner for trees cut from the land, usually contracted in dollars per cord or per M. Stumpage also referred to the right to cut trees on someone else's land.

SUGAR BUSH—A stand of sugar (hard) maple from which sap is drained in the spring to make maple sugar.

TALLY BOY—One who recorded or tallied the measurements of logs as they were called by the scaler.

"TENDING CENTRAL"—Taking and connecting calls for the telephone company, at a time when all calls in or out of an area went through the telephone company's central switchboard.

TIE CUT—A log, usually cedar, cut for use as a railroad tie.

TIMBER CRUISER—Another name for a **LAND-LOOKER**.

TIMBER FITTER—A man whose job was to size up a tree, determine the exact spot where it was to fall and notch it accordingly.

TIME BOY—An office worker who gathered time and purchase slips from the company's mill, camp and yard foremen and brought them to the central office for payroll and account bookkeeping.

TIME CHECK—In the spring, when the camps broke up, the camp clerk gave each of the lumberjacks a time check for their pay. These could be redeemed for cash at the company office or at stores in town. At times, time checks were used as a "paper promise" to pay and were given to jacks in the spring in lieu of cash, redeemable in cash in the fall when the logs were sold. Stores might take a percentage of the cash when they redeemed the time checks during the summer.

When a lumberjack quit his job, the camp clerk gave him a time check showing the number of days he had worked, and he had to take this time check to the head office to get his pay.

TOTE TEAMSTER—Driver who brought supplies into a logging camp.

TURKEY—Packsack or any kind of sack such as a meal sack in which a lumberjack carried his belongings.

UPPER PENINSULA—This term dates from the time of the fur trade. It was upstream from Lake Michigan, and was called *pays d'en haut*, meaning "the upper country."

VAN—The small store in a logging camp in which clothing, tobacco and medicine were kept to supply the crew. Similar to **WANIGAN**.

WALKING BOSS—The superintendent of two or more logging camps. A man elected from among the river drivers to be boss of all the crews on the drive.

WANIGAN—Floating cook shack for feeding men on a river drive. Various spellings, including **WANGAN**, **WANAGAN**. A flat boat carrying a cook stove and supplies which floated down the rivers behind the log drives; it also came to mean the supply shed in the logging camp, a sort of company store that stocked the numerous items a lumberjack needed to carry him through the winter.

WIDOW MAKER—A tree lodged against another tree, or a hanging branch suspended in the top of a tree. A hanging branch could be released by wind or otherwise fall without warning, or would spring back (Springpole) in the felling process.

W.R.C.—Women's Relief Corps. In Manistique, organized April 8, 1892, with fourteen members.

YAWL—A fore-and-aft rigged two-masted vessel similar to a ketch.

YORK STATE—New York State. Many settlers and lumbermen in the Upper Peninsula of Michigan, including the seven Orr brothers, came from New York State.

SOURCES FOR GLOSSARY

Crowe, William S. *Lumberjack*, 2nd Edition. Ted Bays, Editor. Senger Publishing Company: Manistique, MI, 1977.

Crowe, William S, Correspondence to and from L. G. Sorden, July and August 1963.

Crowe, William S. Correspondence to Ferris E. Lewis, December 9, 1952.

Merriam-Webster Collegiate Dictionary, 10th Edition. Merriam-Webster, Inc.: Springfield, MA, 1994.

Orr, Jack. *Lumberjacks and Other Stories*, private publication, Manistique, MI, 1982. By permission of E. Marion Orr.

Paul, Helen Longyear. *Landlooker*, from the reminiscences of John Munro Longyear, Marquette County Historical Society, 1960.

Premo, Claude R. *Dams and Camps on the East Fence River*, private publication, 1972. By permission of Dean and Kent Premo.

Schoolcraft County Historical Society, Manistique, MI.

Sorden, L. G., and Jacque Vallier. *Lumberjack Lingo*, NorthWord, Inc.: Minocqua, WI, 1986.

Sorden, L. G. *Loggers' Words of Yesteryear*, Wisconsin House: Madison, WI, 1956.

Worth, Jean. Personal Communication, Escanaba, MI.

TWO-MAN CROSSCUT SAW

ACKNOWLEDGMENTS

FOR THE THIRD EDITION (2002)

The co-editors of the Fiftieth anniversary edition are indebted to many people and many sources, as we sought to reproduce William Crowe's writings and to clarify information and references which have grown dim in the ensuing fifty years. Where we have used specific written sources, we have referenced those sources in the appropriate Editors' Notes and in the Glossary.

For materials loaned, for enthusiasm shared and for specific information, we thank old and new friends, including relatives of William S. Crowe. They graciously answered questions and shared the results of their own inquiries, as well as memories, documents and photographs:

Ellen McFeeters Turnbull, who supplied both personal memories and copies of family histories written by her and by her father, Milo Cameron McFeeters; **E. Marion Orr** of Manistique, who searched the attic for photos and manuscripts, gave us tea and granted permission to use materials from Jack Orr's "Memories" columns and books; **Rod Smith** of Marquette, MI, for sharing his extensive knowledge of Upper Peninsula log marks and their history; **Vicki Herrmann**, for photos and family memories of early Steuben; the **Premo family** of Amasa, MI—Bette, Dean, Evan and Laurel, for *"The Log Driver's Waltz"* and for a look into the logging history along the Fence River; **Thora Atwater; Margaret Cain; Richard Crowe; Dick Kierzek; Jim Kincade; Marian G. and Ellis Burton Orr; Nina Mattson Orr.**

Our appreciation also goes to those representatives of organizations, historical societies, libraries and businesses who may have been "doing their job" but who made our efforts significantly easier, and helped us find many answers about people and practices "way back then" and now:

The reference staff of the **Peter White Public Library** and of the **Marquette County Historical Society**—J. M. Longyear Research Library; **Vonciel LeDuc**, Schoolcraft County Historical Society, for research, materials and one freezing hour in a snowbound Historical Museum; **Gary Maddox**, Sheriff, Schoolcraft County; **Gus Walker**, Logging Museum, Newberry, MI, for Luce County logging highlights and for also opening up a snowbound Museum; **Lisa Demers**, *Manistique Pioneer-Tribune*, for access to the paper's photo files and permission to reproduce materials; **Mark Harvey**, Reference Archivist, State Archives of Michigan; the on-line and on-site reference librarians at the **State Library of Michigan**; **Rob Berg**, Hartwick Pines State Park; **Mary Jo Remensnyder**, *Michigan History*; **Eric Nordberg**, Archivist, Michigan Technological University, Houghton; **Jack Deo**, Superior View Studio, Marquette; **Robert Sprague**, Porcupine Mountains Wilderness State Park; **Ann Wilson**, Communications Representative, Michigan Department of Natural Resources—Upper Peninsula; **Glen Matthews**, Michigan DNR, Gaylord; **Earl Wolf,** Office of Information & Education, Michigan DNR, Lansing; **Holly Peitsch**, Timber Producers Association of Michigan and Wisconsin; the forestry and lumber men: **Jim Schneider**, Marquette, of the former Schneider Brothers Lumber Company, **Chris Burnett**, Big Creek Forestry, Marquette, **John Hebert**, Sawyer-Stoll

Company, Escanaba, and **Jake Hayrynen**, Longyear Company of Marquette; **Barbara Harold** of the former NorthWord, Inc. for permission to quote from *Lumberjack Lingo*; **David Kuzma**, Special Collections and University Archives, Rutgers University; **Carol Green**, Head, Forest Resource Library, University of Washington, Seattle; **Ben Kudwa**, Michigan Potato Industry Commission; **Chad Cloos**, Michigan Agricultural Statistical Service; **Lisa Hinzman**, Wisconsin Historical Society and **David Tsuneishi**, National Library of Education, Washington, D.C.

Lynn M. Emerick
Ann M. Weller

FOR THE SECOND EDITION (1977)

I hope this full edition carried out my father's hopes for the re-publication of his book, and would like to acknowledge the role of the following people in this endeavor:

Harold "Rabb" Klagstad, for the original publication of *Lumberjack*. I regret that he did not live to see the full edition published.

Ted Bays for his excellent editing and valuable counsel.

Publisher Frank Senger.

Harvey Saunders, for his much appreciated information on river driving..

And to all others who may have contributed information for this publication, much appreciation

Helen Crowe McGlothlin

Such research as natural curiosity provokes utilized the kind assistance and resources of Alex Meron and the Schoolcraft County Historical Museum; Mrs. Katherine LeBrasseur and the Manistique Public Library; Mrs. Helen McGlothlin; James L. Carter, of the Northern Michigan University Press; and the sources noted in the Glossary. Arnold Mackowiak, Frank Senger and the staff of the *Manistique Pioneer-Tribune* and *Delta Reporter* produced the book in a design and style consistent with its great historical value.

Ted L. Bays, Editor

FOR THE FIRST EDITION (1952)

In preparing these chapters, I wish to acknowledge with thanks information received from Harvey Saunders, John I. Bellaire, Mike Kotchon, William Turpin, Everett Cookson, Julius Phillion, Charles Hancock, Charles Slining, Ed Jewell, Alex Creighton, George Leonard, Mrs. E. W. Miller, Ida Maclaurin, Leslie Bouschor, Laura Williams, W. J. Shinar, Charles Robinson, Henry Jahn and Courtney E. Carlson.

William S. Crowe

FRONT COVER: Stutts Creek river drive in the 1890s—W. S. Crowe collection

BACK COVER: World's Fair log load, 1893. Shipped to Chicago as part of the Michigan exhibit at the 1893 World's Fair, the fifty white pine logs in the load were cut six miles south of Ewen, Michigan, in Ontonagon County, hauled to the Ontonagon River and floated to town. Total board feet in the eighteen-foot-long logs was 36,055. The sled, of bird's-eye maple, was built by Wm. Elder; the chains holding the load weighed two thousand pounds and two horses pulled the load eight hundred feet on an iced road. At the fair, lumberjacks showed their skills by unloading and loading the sled each day. Superior View Studio, Marquette, MI

Log end photo courtesy of Dr. Roger Rosentreter, *Michigan History* magazine.

ABOUT THE EDITORS

Lynn M. Emerick and Ann M. Weller spent their early years in Tennessee; in 1948 they moved with their mother to Manistique, a small town on the northern shore of Lake Michigan, where they had spent many summers. Their grandfather, William Crowe, still lived in the family home and was involved in writing his memoirs of life and logging in the late 1890s and early 1900s.

When they were youngsters in Tennessee, grandfather Crowe sent them letters on birchbark (from downed trees) and fanciful stories about animals and life in the Upper Peninsula. His summertime through-the-woods treks near Indian Lake with his grandchildren, to observe nature and to pick blueberries, left lasting memories.

LYNN EMERICK has been a public school speech therapist in the schools of Michigan and Minnesota, an emergency medical services coordinator and the director of a county commission on aging. In 1994, she founded Emerick Consulting, providing publicity, marketing and evaluation services for nonprofit agencies and educational institutions as well for a local book publishing company. She and her husband, Lon, live in a log home in Marquette County, near the south shore of Lake Superior. They enjoy hiking, flyfishing and exploring Upper Peninsula woods and waters, including locales prominent in Crowe's writings about early logging: Stutts Creek, the Indian and Driggs rivers, the Manistique River's High Rollaways and the Haywire railroad grade. They have two daughters working in wilderness and wildland fire management in Alaska and Oregon.

Despite years of woods walks with her grandfather, a summer as resident manager of his Sunset Beach resort and sailing with him—well into his eighties—in a kayak on Indian Lake, she did not fully realize the richness of the life he led in the white pine years until she and her sister Ann developed the Third Edition of *Lumberjack*.

ANN WELLER started her writing career on Manistique High School's student newspaper and later worked for the *Manistique Pioneer-Tribune* one summer when she was in journalism school at Michigan State University. She was also a summer employee of the city's tourist office, a drive-in restaurant, and the Surf restaurant, all of which involved telling stories—most of them true—about Manistique and the Upper Peninsula to travelers.

After working for magazines and publishing companies in Chicago and Detroit and the Library of Congress congressional reference service in Washington, DC, she moved to East Lansing and was employed in state government public information offices for commerce and for housing development. In 1992, she founded Services for Publishers, a free-lance editing and proofreading company. Now living in Holland, MI, on Lake Michigan's east coast, she is involved in cross-cultural and racial justice issues, serves on the board of the Holland Historical Trust, enjoys travel and reading and is learning to speak Spanish and to play the harmonica, though not at the same time. Between them, she and her husband, Herb, have three children, five grandchildren, and one great-grandchild.

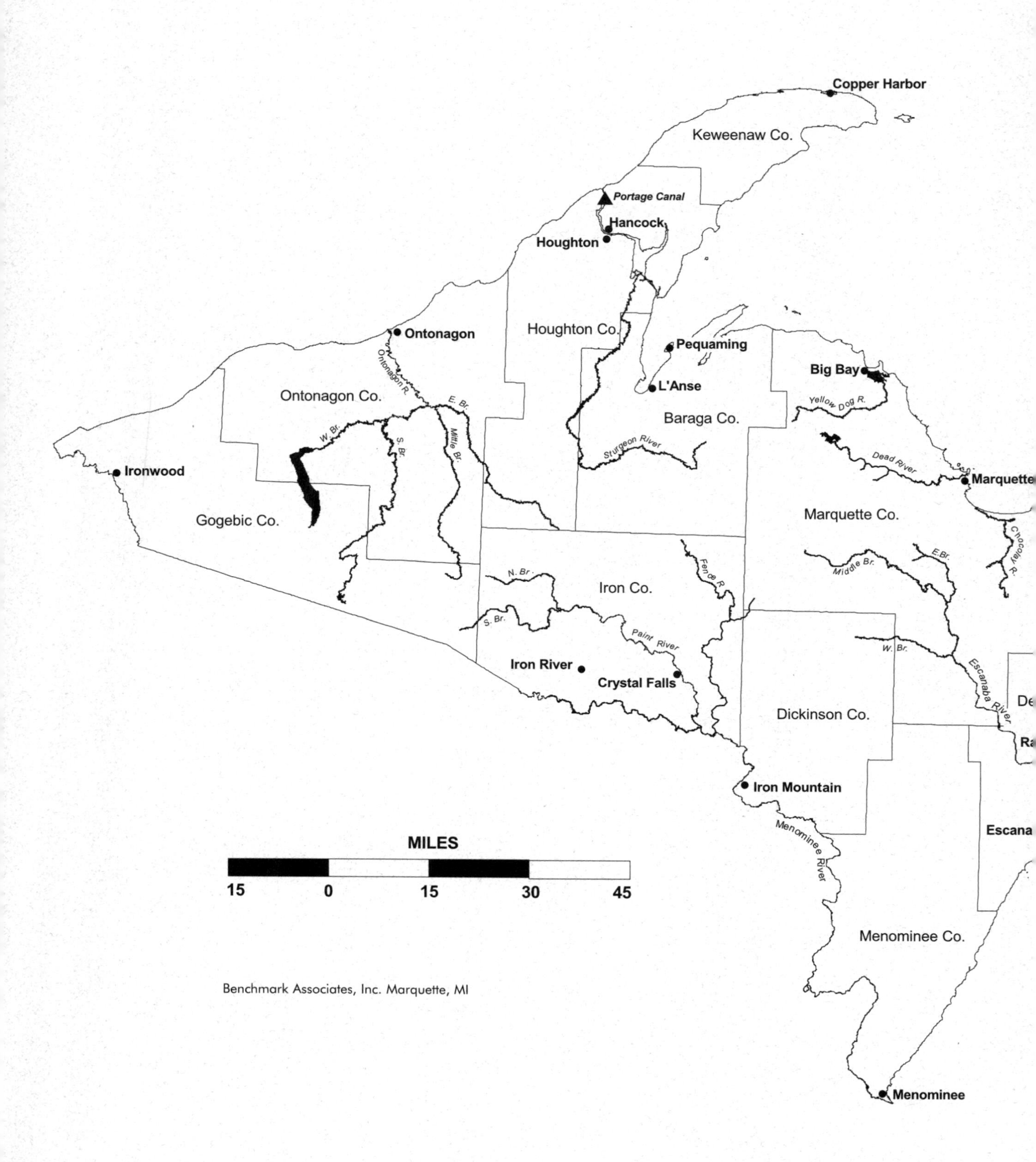

Copper Harbor
Keweenaw Co.
Portage Canal
Hancock
Houghton
Houghton Co.
Ontonagon
Ontonagon R.
Ontonagon Co.
E. Br.
W. Br.
S. Br.
Middle Br.
Ironwood
Gogebic Co.
Pequaming
L'Anse
Baraga Co.
Sturgeon River
Big Bay
Yellow Dog R.
Dead River
Marquette
Marquette Co.
Chocolay R.
E.Br.
Middle Br.
W. Br.
Escanaba River
Fence R.
Iron Co.
N. Br.
S. Br.
Paint River
Iron River
Crystal Falls
Dickinson Co.
Iron Mountain
Menominee River
Menominee Co.
Menominee
MILES
15
0
15
30
45
Benchmark Associates, Inc. Marquette, MI

UPPER PENINSULA - MICHIGAN

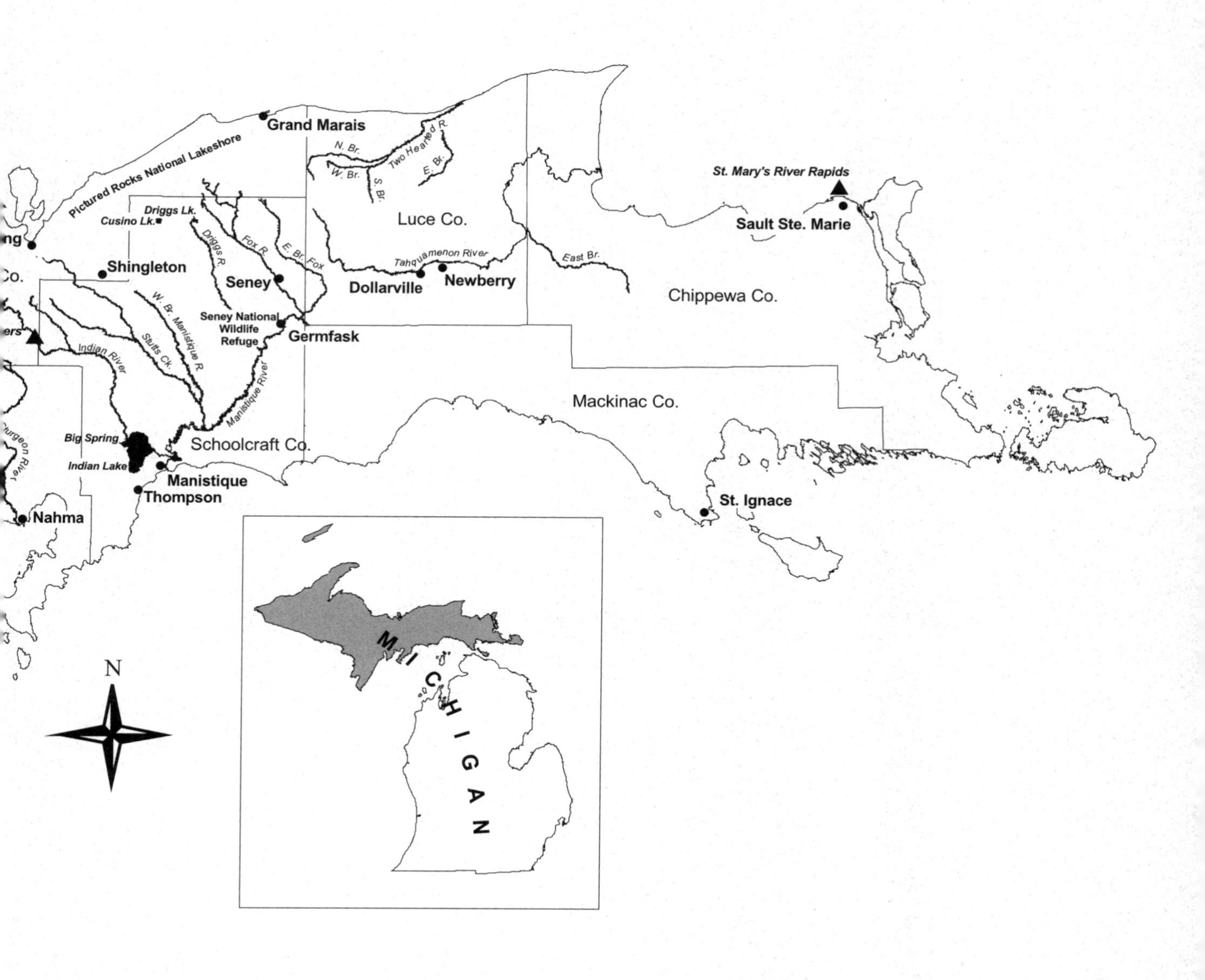